EXAM *Revision* NOTES

AS/A-LEVEL
Biology

Bill Indge

2nd Edition

This book is written for students who started an AS or A-level biology course from September 2008 onwards. It is a revision of the first edition and has been amended to match the new specifications (syllabuses) and examinations for AS/A-level biology from 2008 onwards.

Philip Allan Updates, an imprint of Hodder Education, part of Hachette UK, Market Place, Deddington, Oxfordshire OX15 0SE

Orders

Bookpoint Ltd, 130 Milton Park, Abingdon, Oxfordshire, OX14 4SB
tel: 01235 827720
fax: 01235 400454
e-mail: uk.orders@bookpoint.co.uk

Lines are open 9.00 a.m.–5.00 p.m., Monday to Saturday, with a 24-hour message answering service. You can also order through the Philip Allan Updates website: www.philipallan.co.uk

© Philip Allan Updates 2009
ISBN 978-0-340-95860-5

First printed 2009
Impression number   5   4   3   2
Year   2014   2013   2012   2011   2010   2009

Printed in Spain

Environmental information

Hachette UK's policy is to use papers that are natural, renewable and recyclable products and made from wood grown in sustainable forests. The logging and manufacturing processes are expected to conform to the environmental regulations of the country of origin.

Contents

AS Biology

Chapter 1 Biological molecules

Chapter 2 Cells

Chapter 3 Exchange and transport

Chapter 7 Control systems

Chapter 8 Ecosystems

Introduction

About this book

This book of revision notes has been designed to be as concise as possible. In it you will find all that you need to know in order to answer questions on the core topics in your AS and A2 specifications — and no more! This is why it is an essential revision guide.

This introductory section is in two parts. The first part tells you how to use the book to form a base for your programme of revision. The second part concentrates on examination technique, and looks at the skills you need to turn your hard-earned knowledge into marks.

The book is then organised into two parts, corresponding to AS and A2 biology respectively. Each of these parts is then organised into four chapters as follows:

AS biology
1 Biological molecules
2 Cells
3 Exchange and transport
4 Biodiversity

A2 biology
5 Energy for biological processes
6 Cellular control
7 Control systems
8 Ecosystems

Each of these chapters is subdivided into a number of topics. As you work through the notes, try not to do too much at a time. Ideally, each topic, and the work you do to support the material it contains, should take one revision session.

You will see that the topics contain a number of features to help you in your revision. Each one starts with a list that gives you details of what you must know and what you must be able to do. It is not a complete list, but it should provide you with a good starting point. However, it is important to appreciate that the specifications from the various examining boards differ slightly in the detail they require. It is vital therefore that you look at the specification for your own biology course to see what you will be tested on. You can then cross-check your specification with the list. Add to the list anything that is missing and cross out anything that you do not need to know. You will now have a personalised checklist that can be used to chart your progress.

The remainder of each topic is devoted to covering the basic points you need. Diagrams, tables and text have all been kept to a minimum so that the necessary detail can be learnt more easily. Bold type has been used to highlight technical terms in the specific places where they have been explained.

Finally, the margin notes on many of the pages are points which AS and A2 examiners recognise as frequently creating problems for students. Read them carefully and try to take note of them. If you follow all these pieces of advice, you will probably avoid losing a considerable number of marks.

Revising for AS and A2 biology

We all learn in our own ways. Some of us like to make an early start to the day, others don't produce their best efforts until the evening. Some like to listen to music while they are working, others prefer silence. Revision is an individual matter and no one can tell you how to revise. You need to adopt the approach that works best for you. What follows are seven basic rules for effective revision. However you go about learning your work, they should be useful.

1 You will need a plan of action. If you are following a modular course, you will probably need about 4 weeks to complete your revision of a particular module. If you are taking all of your examinations at the end of the course, you will need much longer. You have to strike a balance between starting too early and risking boredom, and leaving yourself too little time and ending with a last-minute panic. The first job is to decide on the time you need, or have available, and produce a timetable. Bear these points in mind:

- Make sure that you start with those topics that you find most difficult. This will give you both time to get help if necessary and an opportunity to revisit them later.
- Plan realistically. You know what other subjects you are studying. You need to devote ample time to them as well. You will also need to think about your social commitments. There is not much point in planning to revise photosynthesis on your eighteenth birthday, for example.
- The best revision timetable is one that you can stick to — so allow a little flexibility for the unforeseen.

An effective plan is an effective start to effective revision.

2 The last thing you want to do is to arrive at the examination so exhausted that you cannot do yourself justice. You need to allow yourself some time off. If you want to go out on a Saturday night, do so and don't feel guilty. Most students' time can be divided into three: the time they spend working, the time they enjoy themselves, and the time spent flopping about doing nothing much at all. The key is to cut down on the last of these.

Pace yourself, you need a break.

3 Perhaps this situation is familiar to you. You go upstairs with the intention of revising respiration. You wonder if the jeans on the bed will do for Saturday or if they want washing. You look through your CDs to find something to listen to and then wonder who has the one you particularly want. You decide you need a cup of coffee first, then go to the bathroom … and so the evening progresses. You've spent 5 hours working but you have very little to show for it.

It is what you do, not how long you spend doing it that matters. Few people can work for more than an hour at a time and keep their minds on the task in hand. It is much better to work in short bursts and be absolutely single-minded. If you find your thoughts wandering, bring them back immediately. At the end of an hour, or half-hour for that matter, give yourself a break. This is the time to get yourself a cup of coffee, or sort those CDs out.

It is quality not quantity that is important.

4 Your revision will never be particularly effective if you don't set yourself realistic targets. You cannot hope to revise all of biology in a single night. It is much better to divide your work into smaller topics, as has been done for you in these revision notes. It always helps if you can look back at the end of a revision session and say that you have covered a particular piece of biology and can tick it off your list.

Realistic targets make good incentives.

5 Revising for an AS or A2 examination is very similar to preparing for your driving test. There are things you need to learn and skills you need to practise. You are not likely to pass your driving test if you haven't learnt about stopping distances or practised hill starts. AS and A2 exams are much the same, except that before you get started you have to understand the basics. In your revision, make sure that you:
- Understand the main ideas. If you have any doubts, ask for help. If you understand the basic principles, it's much easier to learn the necessary detail.
- Learn the facts.
- Practise the skills. Once you have completed the first two steps, find some questions and make sure that you can turn your knowledge into marks. If you have problems, get some help.

Understand the principles, learn the facts and practise the skills.

6 It is very easy to sit admiring your notes and hope that information will simply diffuse from the paper into your brain. Get a pen and paper and make use of them. Write down the points that you have forgotten. Test yourself.

Make revision an active process.

7 You will never get anywhere if you don't get started. You may wish to spend time rearranging your notes, decorating timetables and reading revision guides, but you won't learn any biology. It is very easy to come up with excuses and put off revision … I need a rest after a long term; I'll start on Monday; 4 weeks will be about right.

The best time to start your revision is now.

Examinations and examination technique

When the AS and A2 results are published every August, there are always some students who have excelled. Unfortunately, there are also always some who have failed to gain the grades they required. They have worked hard throughout the course and have a sound grasp of their basic biology. What has gone wrong? It may come as something of a surprise, but it is frequently not a lack of knowledge that leads to disappointing results, but poor examination technique resulting in the inability to turn facts into examination marks. What can be done about it?

The first step in the right direction is to make sure that you follow the advice given in the previous section and practise the necessary skills. Try this overall approach:

- Find the topic that you are going to revise in these revision notes. Look at the list at the beginning of the topic together with your specification to produce your personal checklist (see 'About this book' on p. 1).
- Use this book to learn the facts.
- While these facts are still fresh in your mind, practise the skills by looking at some past examination questions. This will enable you to get into good habits and make sure that you can turn your knowledge into marks. If you don't have access to any past papers, you can buy them from your examination board's publications department. The address will be in the specification.

Before looking at individual questions, it is worth thinking about how AS and A2 questions should be answered. Whatever the question, there are three basic steps in the process which we can summarise as a flow chart:

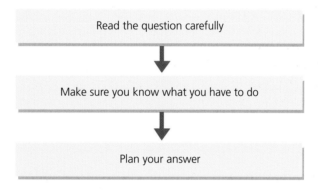

We will look at each of these in turn.

Reading the question

It should be obvious that you will only get marks if you answer the question that has been asked, not the one that you thought was asked. Examiners try to make the wording of questions as straightforward as possible but, in the stress of examination conditions, it is all too easy to misinterpret a question. Take this question from a recent paper:

> *The size of a population of woodlice may be estimated by using the mark-release-recapture method. What would you need to measure in order to estimate the woodlouse population by this method?*

The key phrase is 'need to measure'. If you have already studied the topic of ecology, you should appreciate that you will need three pieces of information:
- the number of woodlice marked and released
- the number of woodlice caught in a second sample
- the number of marked woodlice in the second sample

An answer that incorporated these points would gain full marks.

However, when the question was marked, the examiners were surprised to see large numbers of answers which contained points like these:
- old tin
- white paint
- paint brush

Candidates had misread the question as:

What would you need in order to estimate the woodlouse population by this method?

A simple mistake, but 3 marks lost.

Read each line of each question carefully.

Reading the question carefully is one thing; understanding it is another. When you have to answer questions which test your ability to interpret information in the form of graphs and tables, it is very important that you get a clear understanding of the data with which you have been provided. This is particularly true if the data are unfamiliar.

Think for a minute about the way in which you learn. If you have revised your work carefully, it is almost as though you have filed away in the relevant part of your brain a stack of cards, each one containing all the bits of information you know about a particular topic. To answer a question well, you need to be able to extract this card. The first step, then, is to read quickly through the question and see what it is about. You might say to yourself, for example, that this is an enzyme question — so you can start thinking in terms of proteins, active sites and collisions between molecules. This gives you a lot of familiar biology on which you can draw.

With unfamiliar data, identify the topic on which the question is based.

Where the individual parts of a question are based on information provided, you need to take time to understand tables and graphs. Look at Figure 1. It shows a graph taken from a question about enzymes. At first sight, it might look familiar.

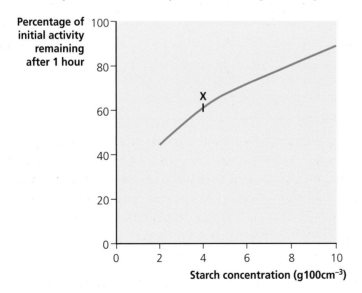

Figure 1

The shape of the curve is very like that obtained when the rate of reaction of an enzyme is plotted against substrate concentration. But it is not! You need to look at the graph carefully to see what the two axes represent. The x-axis is straightforward enough. It shows the concentration of the substrate, starch. The y-axis, however, needs a little more

thought. It compares the activity of the enzyme after 1 hour with its activity at the start. In other words, it tells you how much of its activity remained after 1 hour in a particular set of conditions. In setting this question, the examiners were not trying to trick candidates, but they did require them to look at the data carefully.

There is a second step that you can take that should give you an even clearer picture. Pick a point on the graph and explain to yourself exactly what it represents. Try it here. Look at the point marked **X** on the curve. What it tells us is that after 1 hour at a starch concentration of 4g 100 cm^{-3}, the amylase would have 60% of the activity that it had at the start of the experiment. To summarise:

- Read the axes of the graph first so that you know exactly what is being plotted.
- Make sure you understand what the graph shows by explaining what one particular point represents.

You can use the same basic approach with a table. Look at Table 1 which forms part of a question taken from an ecology module test.

Table 1 shows the results of an investigation into how the distribution of the roots of three species of grass varied with depth. The figures are the percentages of the total root dry mass of the species concerned.

Soil depth (m)	Species of grass		
	Panicum maximum	*Themedra triandra*	*Eragrostis superba*
0–0.4	64.9	66.5	73.6
0.4–0.8	14.2	25.9	15.5
0.8–1.2	12.1	5.6	7.4
1.2–1.6	4.7	1.4	2.6
1.6–2.0	2.6	0.6	0.8
2.0–2.4	1.2	0	0.1
2.4–2.8	0.3	0	0
Total dry mass (g per plant)	114.0	58.3	27.2

Table 1 Depth of roots in various grass species

Read the accompanying question carefully, and then read the table headings. If you are confident that you understand what is shown, test yourself by describing as accurately as you can what the shaded box represents. If you don't take care, you can go wrong, particularly when you are faced with the stress of an examination. To illustrate the point, here are two comments made by candidates answering this question:

- Most *Eragrostis* plants grow in soil less than 0.4 m deep.
- *Eragrostis* is the commonest plant in shallow soil.

Neither of these comments describes the information in the shaded box. If you have understood the information, you should be able to see that this box shows that 5.6% of the total dry mass of the roots of the grass, *Themedra triandra* are found at a depth of between 0.8 and 1.2 metres.

Take time to understand information in tables and graphs.

Making sure you know what you have to do

Once you have read the question and understood the information with which you have been provided, you can look in more detail at what you are required to do. AS and A2 questions come in a variety of different forms, but if you look at a large enough number you will begin to recognise a pattern. Part of the question will provide you with information. This may be a short paragraph, a diagram, a table or a graph. The other component will provide you with instructions as to what the examiner wants you to do. It is this that we will now look at in a little more detail.

Examiners think very carefully about the instructions they use in their question papers. It is very much in their best interest to provide clear directions that can only be interpreted in one way. It is obviously equally important that you are familiar with the terms that are used and respond in the right way. The following list contains some of the common instructions used in biology examinations. It doesn't contain all the terms used because, if examiners can think of a way of making a question even clearer, they will do so, even if it means using a word that is not on this list. In addition, obvious question words such as 'When?', 'Where?' and 'How?' have been left out. They are unlikely to cause difficulties. You should compare this list with the questions set by the examination board whose papers you are taking. If they use other instructions, add them and write your own meaning.

Some common instructions used in AS and A2 biology exams

Calculate

This term doesn't really need an explanation but there are two points that are worth making. Make sure that you show your working as clearly as possible. Then, even if you make an arithmetical mistake, you may still get some credit. You won't get any marks, however, if the examiner cannot understand or follow what you have done. The second point is to make sure that you have given the units in which your answer is expressed.

Compare

Describe the similarities and differences between the features concerned. Make sure that you don't simply describe one feature and leave the examiner to do your work for you. It is worth answering questions which require comparison in the form, 'one has ... while the other has ...'

Describe

This instruction is often related to material presented in a table or graph. In this case, you need to write down what can be seen in terms of the general pattern or trend in the figures. The term is also used in longer questions such as 'Describe how haemoglobin loads oxygen in the lungs and unloads it in a muscle cell'. In this case, you will need to give a step-by-step account of what happens, using the mark allocation as a guide to the amount of detail you need.

Evaluate

This involves explaining the evidence for the points you are trying to make.

Explain

Explain is probably the most frequently misinterpreted instruction of all. It requires you to give a reason. One of the best ways of checking that you have done what is required is to ask yourself 'Have I said *why*?' An explanation is not the same as a description.

Name

Name usually requires no more than a single-word answer. Don't waste your time by writing out the question first. 'The name of the DNA base that is complementary to adenine is thymine' will take a lot longer to write down than 'thymine' but it will get the same number of marks.

Sketch

The instruction used when a curve has to be drawn on a graph but where you can't be expected to know the exact figures. All that is necessary is an indication of the shape of the curve. You don't need to invent figures and spend time plotting these accurately.

State

This is really a way of telling you that all that is needed is a short, concise answer. Some examination boards use the term 'give' instead.

Suggest

There are two uses for this term. If there are several valid answers to a particular question, then the examiner may ask you to suggest an answer. It is also used in cases where you are not expected to know the precise answer but should have enough biological knowledge to be able to offer a sensible idea.

Use information in...

This involves making use of particular information to illustrate your answer. Since this is a requirement of the question, you are unlikely to get many marks for a general answer that does not draw on the necessary information.

Now you have a list of instructions, use it! If you are at the beginning of your AS or A2 course, you could make a copy of this list and keep it by you when you are answering questions. Check constantly that you are doing what you are supposed to be doing. The aim is to get so familiar with these instructions that you will always make the right response. Above all:

When you are asked to explain, make sure you give a reason why, and when you are required to use a particular piece of information, use it!

Planning your answer

In the examination which you will be taking, there will be a number of questions which are broken down into parts, each part worth something like 4 or 5 marks. Here are two examples:

Explain the part played by haemoglobin in transporting oxygen from the lungs to respiring muscle cells. (6 marks)

Explain how ATP is produced during the light-dependent reactions of photosynthesis. (4 marks)

Many candidates find difficulties with this sort of question. They fail to do themselves justice even though they understand the biology involved.

This is where planning your answer becomes important. Planning doesn't mean writing out an elaborate outline; all that is required is an indication of the points you need to make in order to get the marks. Once you have revised a particular topic, find some questions of this type and have a go at them. Don't waste time writing out answers in full. Look at each question carefully and list the points you need to make in order to gain full marks. Concentrate on each question fully, just as you would in an examination, and work quickly. You need to put yourself under the same sort of pressures as you will encounter in the examination.

Check your answers against this checklist:
- Have you included enough points? You need to make a valid point for each mark in the mark allocation.
- Have you answered in the right level of detail? Each point that you make should be based on material that you have been taught in your AS or A2 course. Remember, you need A-level information to earn marks.
- Have you selected relevant material? Check your answer again and make sure that you have answered the question. It is much easier to change a plan than change the final answer.

Plan what you intend to write. Never forget that one valid A-level point is worth one mark.

AS
Biology

Biological molecules

Big molecules and small molecules

When you have finished revising this topic, you should:

- know that many important biological molecules are polymers
- be able to explain how polymers are formed by condensation of small monomers, which are based on a small number of chemical elements
- know how proteins, carbohydrates and lipids may be identified by simple biochemical tests
- be able to explain how chromatography can be used to separate and identify substances present in a mixture

1.1 Monomers and polymers

A **polymer** is a large molecule formed from many identical or similar smaller molecules called **monomers**. These monomers are joined by condensation. **Condensation** is a chemical reaction which involves removing a molecule of water. When a large number of glucose molecules are joined by condensation, a molecule of starch is formed. Glucose is a monomer and starch is a polymer.

Just as monomers can be joined by removing molecules of water, polymers can be broken down by adding molecules of water. This type of reaction is known as **hydrolysis**. Figure 1.1 summarises condensation and hydrolysis.

> Don't confuse condensation and hydrolysis. Condensation involves joining together with the removal of water; hydrolysis is breaking apart with the addition of water.

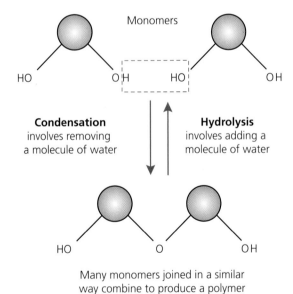

Figure 1.1 *Condensation and hydrolysis*

Some important biological molecules, such as lipids, are formed by condensation of smaller molecules, but they are not polymers, because they are not built up from a large number of similar monomers. Because of this, it is useful to use the word **macromolecule** for all large, biologically important molecules. Table 1.1 summarises the properties of the biologically important macro molecules with which you should be familiar.

Macromolecule	Basic building blocks	Elements present	Main functions in living organisms	Further details
Polysaccharides – starch, glycogen and cellulose	Monosaccharides	Carbon, hydrogen and oxygen.	Starch and glycogen are storage compounds. They can be broken down to form glucose, important in respiration. Cellulose is an important component of plant cell walls.	See Chapter 1, Section 2.
Protein	Amino acid	Carbon, hydrogen, oxygen and nitrogen. Many proteins also contain sulphur.	Proteins have many functions. They act as receptors, they form antibodies, control reactions as enzymes and are important structural substances in tissues such as muscle and bone.	See Chapter 1, Section 3.
Nucleic acids – DNA and RNA	Nucleotides	Carbon, hydrogen, oxygen, phosphorus and nitrogen.	Nucleic acids carry the genetic code of an organism and are involved in protein synthesis.	See Chapter 6, Section 1.
Lipids – triglycerides (fats and oils)	Fatty acids and glycerol	Carbon, hydrogen and oxygen.	Important storage compounds and can be used for respiration.	See Chapter 1, Section 4.

Table 1.1 Some biologically important macromolecules

1.2 Biochemical tests

Table 1.2 contains a list of the biochemical tests with which you should be familiar. Make sure that you know how to carry out each test and the result that you would expect.

1.3 Chromatography

Chromatography is an important way of separating the various substances present in a mixture. The technique usually used in AS and A2 investigations is paper chromatography. The method is summarised in Figure 1.2.

When you have run a chromatogram in this way and separated the mixture into its components, it is useful to be able to identify these substances. To do this you need to calculate the **R_f value** of the substance concerned. This is given by the equation:

$$R_f = \frac{\text{distance moved by the spot (usually measured to its front edge)}}{\text{distance moved by the solvent}}$$

The R_f value of a particular substance will always be the same, provided that the same solvent is used.

Biochemical tests are very basic, but it is surprising how many candidates don't know them.

In answering questions on chromatography, do not confuse the terms 'solute', 'solvent' and 'solution'.

Substance	Test	Brief details of test	Positive result
Protein	Biuret test	Add sodium hydroxide solution to the test sample. Add a few drops of dilute copper sulphate solution.	Solution turns mauve.
Carbohydrates Reducing sugars	Benedict's test	Heat test sample with Benedict's reagent.	Orange-red precipitate is formed.
Non-reducing sugars		Check that there is no reducing sugar present by heating with Benedict's reagent. Hydrolyse by heating with dilute hydrochloric acid. This will convert the non- reducing sugar into reducing sugars. Neutralise by adding sodium hydrogencarbonate. Test sample by heating with Benedict's reagent.	Orange-red precipitate is formed.
Starch	Iodine test	Add iodine solution.	Turns blue-black.
Cellulose		Add Schultze's solution.	Turns purple.
Lipid	Emulsion test	Dissolve the test sample by shaking with ethanol. Pour the resulting solution into water in a test tube.	A white emulsion is formed.

Table 1.2 Biochemical tests

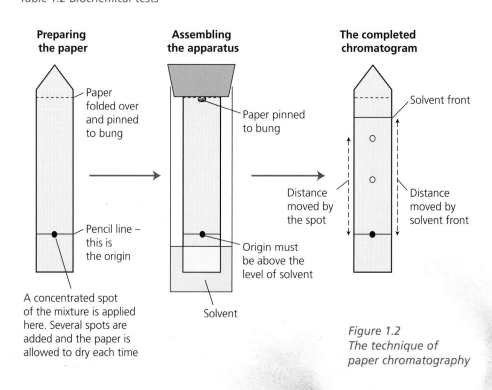

Figure 1.2
The technique of
paper chromatography

2 *Carbohydrates*

When you have finished revising this topic, you should be able to:

- list the general properties of carbohydrates
- draw simple diagrams to show the structures of α-glucose and β-glucose
- show how α-glucose molecules are linked by condensation to form maltose, starch and glycogen
- show how β-glucose molecules are linked to form cellulose

2.1 Introduction

Carbohydrates contain the elements carbon, hydrogen and oxygen. As the hydrogen and oxygen atoms are present in a two to one ratio, we can write the general formula of a carbohydrate as $C_x(H_2O)_y$. Carbohydrates are conveniently classified as **monosaccharides**, **disaccharides** or **polysaccharides** according to whether they contain one, two or many sugar units. In addition, it is possible to have sugar or saccharide units which have different numbers of carbon atoms. **Triose** molecules have three carbon atoms, **pentoses** have five carbon atoms and **hexoses** have six carbon atoms. In general, monosaccharides and disaccharides are soluble, while the larger polysaccharides are insoluble. Table 1.3 summarises the structure of some biologically important carbohydrates.

Monosaccharides	Disaccharides	Polysaccharides
Trioses triose		
Pentoses ribulose ribose deoxyribose		
Hexoses glucose fructose galactose	Maltose (glucose + glucose) Sucrose (glucose + fructose) Lactose (glucose + galactose)	Starch Glycogen Cellulose

Table 1.3 Some biologically important carbohydrates

2.2 The structure of α-glucose

Glucose is a hexose and, as such, each molecule has the formula, $C_6H_{12}O_6$. Unfortunately, this does not show us how the atoms are arranged. To a biologist, the most useful way of representing an α-glucose molecule is that shown in Figure 1.3.

Maltose, sucrose and lactose are carbohydrates. Maltase, sucrase and lactase are enzymes that hydrolyse these disaccharides. Make sure that you write clearly and distinguish between the carbohydrates and the enzymes.

All students know that a glucose molecule contains six carbon atoms ... until they are asked in a question!

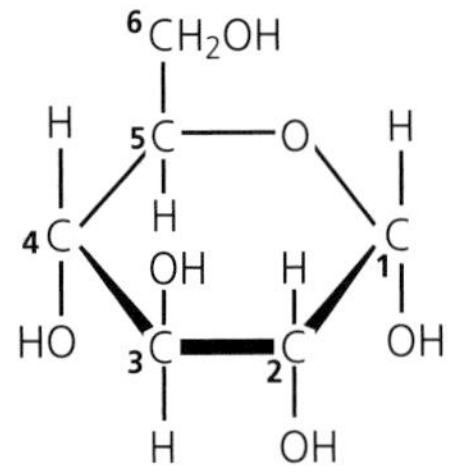

This shows glucose:

- As a ring-shaped molecule with five carbon atoms and an oxygen atom forming the ring.
- With thickened lines which give some idea of the three-dimensional shape of the molecule.
- As having carbon atoms that are numbered 1–6. This is useful if we need to refer to individual atoms.

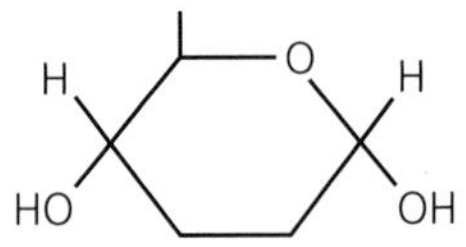

But… this is all you need to learn. It is a simplified version that is adequate for A-level purposes.

Figure 1.3 The structure of an α-glucose molecule

2.3 Maltose and starch

Two α-glucose molecules may be joined together by condensation. A molecule of water is lost and a disaccharide called **maltose** is formed. The bond formed between these two sugar units or **residues** is called a 1–4 **glycosidic bond**, as it joins carbon 1 of one glucose residue to carbon 4 of the other. If we join a large number of these α-glucose molecules together we get a molecule of **amylose**, an important component of starch (see Figure 1.4).

This is only a small part of an amylose molecule. Amylose is made up of many glucose units.

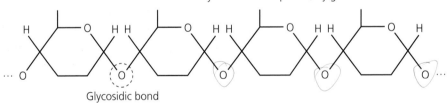

Glycosidic bond

Figure 1.4 The structure of amylose, an important component of starch

The molecules of amylose coil into a helix. Hydrogen bonds hold this helix in shape. Starch is a mixture of amylose and amylopectin. **Amylopectin** is also formed from α-glucose, but its molecules are branched chains, because some of the glycosidic bonds form between carbon 1 and carbon 6. There are many different forms of starch. Corn starch, for example, is different from potato starch. Starches differ in the proportion of amylose and amylopectin that they contain. Starch is a storage molecule and shows a number of features which make it ideal for this purpose:

- It is compact in shape, so a large amount of starch can be packed into a relatively small volume.
- It is insoluble and will not affect the water potential of the cells in which it is stored.

● It is readily hydrolysed to give glucose, which can be used as a respiratory substrate.

Glycogen is another polysaccharide formed by condensation of α-glucose. It is very like amylopectin in its chemical structure and also consists of branched chains.

2.4 Cellulose

β-glucose molecules (see Figure 1.5) can also be linked to each other by 1–4 glycosidic bonds. To bring the OH groups into the right position for a bond to form, we need to flip one of the β-glucose molecules over. In the diagram (Figure 1.6), this molecule appears upside down. If we join a large number of these β-glucose molecules together, we get a molecule of cellulose. Cellulose molecules form long, straight chains. Hydrogen bonds form between neighbouring chains.

Figure 1.5 The structure of a β-glucose molecule

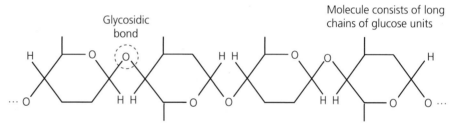

Figure 1.6 The structure of cellulose

Cellulose is an important component of plant cell walls. The features that it shows which make it ideal for this purpose include:
● The hydrogen bonds which link one chain to another produce rigid cross-links which bind the chains into bundles called **microfibrils**. These microfibrils have considerable tensile strength, which means that they are particularly resistant to pulling forces.
● Cellulase enzymes are rare in nature, so relatively few organisms are able to digest plant cell walls.

3 Proteins

When you have finished revising this topic, you should:

■ be able to draw a simple diagram showing the general structure of an amino acid molecule

■ be able to show how amino acid molecules can be joined by condensation to produce dipeptides and polypeptides

■ be able to explain what is meant by the primary, secondary, tertiary and quaternary structure of a protein

■ understand that the tertiary structure of a protein gives the molecule a specific shape — this shape explains many of the functions of proteins in a living organism

3.1 Introduction

The basic building blocks of proteins are **amino acids**. A protein, therefore, contains the same chemical elements as those found in amino acids. Carbon, hydrogen, oxygen and nitrogen are always present. Most proteins also contain some sulphur. Because the different amino acids may be combined in many different ways, there are many different proteins with different functions:

- Proteins are involved in support and movement. In a mammal, for example, proteins form an important component of bone and muscle.
- Enzymes are proteins. Enzymes enable biochemical reactions to take place rapidly under the conditions which exist in living organisms.
- Protein molecules are important components of cell membranes. One of their functions here is to help to control the movement of substances into and out of cells.

DNA is a nucleic acid. It is not a protein. DNA is made up from nucleotides, not amino acids.

3.2 Amino acids

There are 20 different amino acids that are used by living organisms to form proteins. All 20 have the basic structure shown in Figure 1.7.

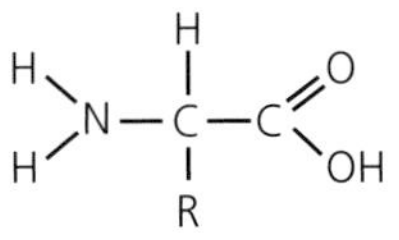

Figure 1.7 The basic structure of an amino acid

All have a central carbon atom to which is attached a nitrogen-containing **amino group** (NH_2), a **hydrogen atom** (H) and a **carboxyl group** (COOH). The remaining group is referred to as the **R group** and differs in different amino acids. In alanine, for example, it is CH_3. In some amino acids, such as cysteine, the R group contains an atom of sulphur.

Make sure that you can draw a diagram of the structure of an amino acid molecule. It is the basic starting-point for writing about protein structure.

3.3 Dipeptides and polypeptides

Figure 1.8 shows how we can join two amino acid molecules together by condensation.

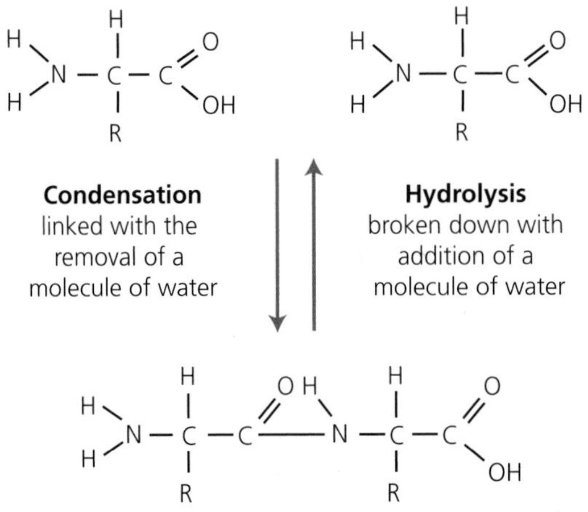

Figure 1.8 Joining amino acids

Condensation
linked with the removal of a molecule of water

Hydrolysis
broken down with addition of a molecule of water

This is a dipeptide.
The two amino acids are joined by a peptide bond.

The bond which joins the two amino acids (or, more correctly, residues, since parts of the molecules have been removed) is a **peptide bond**. If two amino acid molecules are joined together, we have a **dipeptide**; if many amino acids are joined in this way, we get a **polypeptide**. Polypeptides may be folded in particular ways to form proteins. The way in which a particular polypeptide folds, is determined by the sequence of amino acids from which it is made up. This determines the position of the chemical bonds which maintain its shape. The sequence of amino acids which makes up the polypeptide is known as the **primary structure** of the protein. The polypeptide may be folded to give rise to the **secondary** and **tertiary structure**. If there is more than one polypeptide chain in a protein, then the protein is described as having a **quaternary structure**. The main features of primary, secondary, tertiary and quaternary structure are summarised in Figure 1.9.

Primary structure

The sequence of amino acids which makes up the polypeptide chain.

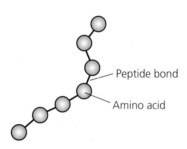

- Proteins differ from each other because their primary structures are different.
- The amino acids are joined by peptide bonds.

Secondary structure

The way in which a polypeptide chain is coiled into a helix or folded in pleats. The diagram shows the coiling of a polypeptide chain to form a helix.

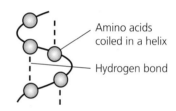

- A particular protein may have a secondary structure which is coiled into a helix, folded into pleats, both or neither of these.
- The secondary structure is maintained by hydrogen bonds.

Tertiary structure

The irregular folding of the polypeptide chain which gives many proteins their globular, three-dimensional shape.

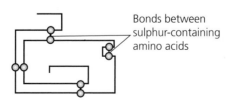

The tertiary structure of a protein is held by chemical bonds. These include:
- hydrogen bonds
- bonds between amino acids that contain sulphur
- ionic bonds

Quaternary structure

A protein has a quaternary structure if it consists of more than one polypeptide chain.

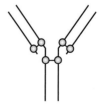

An antibody. This is made up from four polypeptide chains.

Figure 1.9 The structure of protein molecules

All molecules have a three-dimensional shape, so it is not enough to define the tertiary structure of a protein as 'its three-dimensional shape'. You should refer to the irregular folding of the polypeptide which gives the protein its globular shape.

The tertiary structure of the molecule provides a protein with a specific shape, and this shape is often linked to the function of the protein. For example, it explains many of the properties of enzymes and it helps to suggest why proteins are particularly important in controlling the movement of substances into and out of cells.

4 *Lipids*

When you have finished revising this topic, you should be able to:

- list the general properties of lipids
- draw simple diagrams representing the structure of glycerol and fatty acid molecules and distinguish between saturated and unsaturated fatty acids
- show how a molecule of glycerol and three fatty acid molecules are linked by condensation to form a triglyceride
- describe how the structure of a phospholipid differs from that of a tri-glyceride

4.1 Introduction

The lipids are a diverse group of compounds that share the property of being soluble in organic compounds but not in water. They include:

- **Triglycerides.** These are often referred to as fats and oils and form important energy reserves. A given mass of lipid will release a large amount of energy when it is oxidised during respiration. Triglycerides also have insulating properties and many mammals living in colder areas have a thick layer of fat under the skin.
- **Phospholipids.** These play an important role in cell membranes.
- **Steroids.** These include cholesterol and the sex hormones, testosterone, oestrogen and progesterone.

4.2 Triglycerides

A triglyceride is made up from a molecule of **glycerol** and three **fatty acid** molecules. The basic structures of these molecules are shown in Figures 1.10 and 1.11.

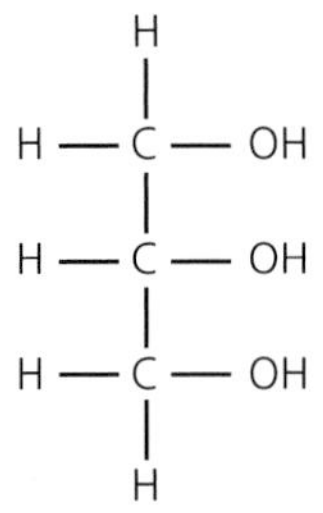

Figure 1.10
The structure of glycerol

Glycerol is a type of alcohol. It has three OH groups, each of which can condense with a fatty acid.

It is easy to learn the structure of a glycerol molecule: three carbons, three OH groups and all the rest are hydrogens.

R.COOH

This is the simplest formula for a fatty acid molecule. The letter R represents a hydrocarbon chain consisting of carbon and hydrogen atoms.

In saturated fatty acids, each of the carbon atoms in this chain, with the exception of the last, has two hydrogen atoms joined to it.

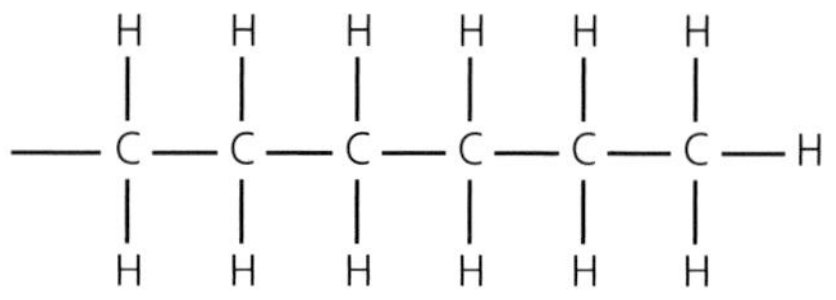

```
    H   H   H   H   H   H
    |   |   |   |   |   |
  — C — C — C — C — C — C — H
    |   |   |   |   |   |
    H   H   H   H   H   H
```

In unsaturated fatty acids, there are one or more double bonds between the carbon atoms in the chain. Because of this, some carbon atoms will be joined only to a single hydrogen atom.

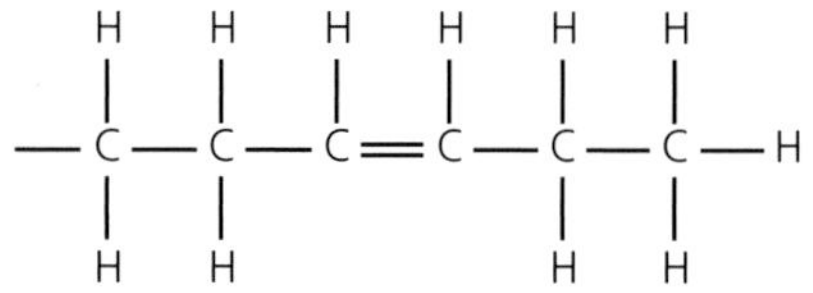

```
    H   H   H   H   H   H
    |   |   |   |   |   |
  — C — C — C = C — C — C — H
    |   |           |   |
    H   H           H   H
```

An unsaturated fatty acid with a single double bond can be described as monounsaturated; an unsaturated fatty acid with more than one double bond in its hydrogen chain is referred to as polyunsaturated.

Figure 1.11 The structure of fatty acid molecules

The fatty acids and glycerol are joined together by condensation to form a triglyceride. The simplest way of representing this is shown in Figure 1.12.

- Draw a diagram to show a glycerol molecule.
- Draw three fatty acid molecules 'the wrong way round' next to it.

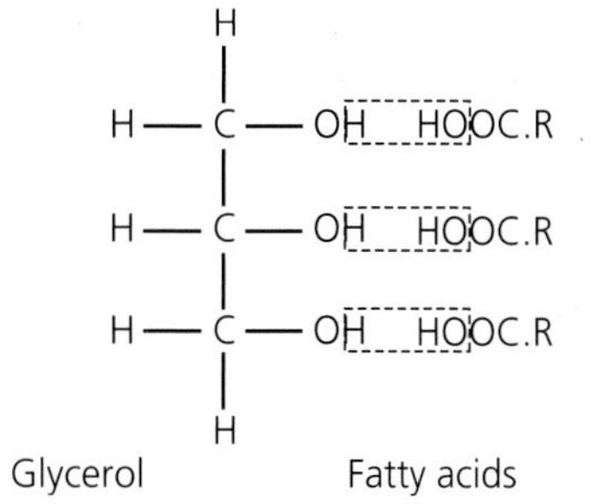

```
          H
          |
  H — C — OH    HOOC.R
          |
  H — C — OH    HOOC.R
          |
  H — C — OH    HOOC.R
          |
          H
  Glycerol      Fatty acids
```

- Remove three molecules of water, taking the H from the glycerol and the OH from the fatty acids.

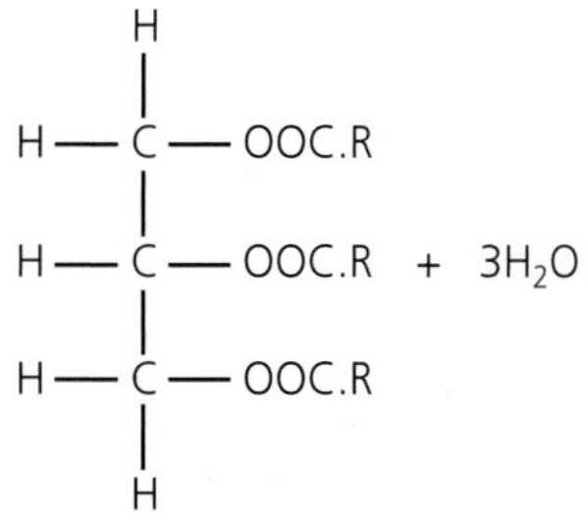

```
          H
          |
  H — C — OOC.R
          |
  H — C — OOC.R   +  3H_2O
          |
  H — C — OOC.R
          |
          H
```

- Close everything up to show the completed triglyceride.

Figure 1.12 Forming a triglyceride

The three fatty acids in a triglyceride molecule may be identical, but they usually differ from one another.

4.3 Phospholipids

A phospholipid is, as its name suggests, a lipid that contains a phosphate group. In a phospholipid, the glycerol joins with two fatty acid molecules and a phosphate group. The phosphate group is water soluble, or **hydrophilic**. The hydrocarbon tails of the fatty acids are insoluble in water and are described as being **hydrophobic**. This property of phospholipids is very important in the structure of cell membranes.

5 *Water*

When you have finished revising this topic, you should:

- be able to explain how hydrogen bonding gives water a number of its properties
- understand why water is an effective solvent and be able to describe its importance in transport systems
- be able to explain the thermal properties of water

5.1 Water molecules and hydrogen bonding

A molecule of water consists of two atoms of hydrogen and an atom of oxygen. The electrons associated with these atoms are not distributed evenly; they tend to be drawn towards the oxygen atom. An electron is negatively charged so this results in the oxygen atom also having a slight negative charge (written as $2\delta^-$) while the hydrogen atoms are left with a slight positive charge (δ^+). Because of this, opposite charges on neighbouring water molecules tend to attract each other, forming **hydrogen bonds** (Figure 1.13). Water molecules are therefore difficult to separate from each other.

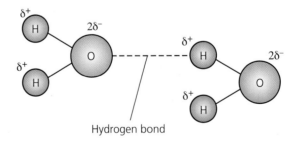

Figure 1.13 Hydrogen bond formed between neighbouring water molecules

5.2 Water as a solvent

Molecules such as those of amino acids and glucose are called **polar molecules** because they have an unequal distribution of charge. Water molecules are not only attracted to each other, they are also attracted to ions and to polar molecules. As a result, molecules and ions of many substances are effectively separated from each other by a skin of water molecules to form a solution. Water

is therefore a very good solvent, and this has two important consequences in living organisms:

1 Substances which are able to move freely in solution have a better chance of colliding with each other and reacting. Water is therefore an excellent medium for biochemical reactions in cells.

2 Diffusion is a very slow process. Large organisms need a more rapid way of moving substances from one place to another. This involves transport systems such as the blood and lymphatic systems of a mammal and xylem and phloem in plants. These systems move substances in solution.

5.3 Cohesion

Because water molecules are held together by hydrogen bonds, they tend to stick to each other. This property of water is called **cohesion**, and it means that it is possible to pull a column of water along a tube. This is the principle that underpins the way in which xylem transports water from the roots of a tree to its leaves.

5.4 The thermal properties of water

Temperature affects water in a different way from that in which it affects many other substances. This provides water with a number of unique thermal properties. Many of these properties are associated with the hydrogen bonds that form between water molecules:

- Hydrogen bonds limit the movement of water molecules, so a relatively large amount of energy is necessary to raise the temperature of a given mass of water. Put another way, the temperature of large amounts of water remain remarkably stable. This is very important for living organisms.

- A large amount of energy is necessary to change water from a liquid to a gas. It is this property that makes the evaporation of sweat particularly effective in cooling the body of a mammal.

- Water has a further unique property — the density of its solid form is less than that of its liquid form. Consequently, an area of water freezes from the top downwards. In practice, this means that a layer of liquid water will remain under the ice of a frozen pond or area of sea in which living organisms can survive extreme weather conditions.

6 *Enzymes*

When you have finished revising this topic, you should:

- be able to explain how enzymes speed up reactions by lowering the activation energy necessary to start the reaction

- know that an enzyme is a protein and be able to explain how its tertiary structure enables an enzyme–substrate complex to be formed

- understand why enzymes are specific and can only catalyse particular intracellular and extracellular reactions

- be able to explain the importance of enzymes in living organisms

- be able to relate the properties of enzymes to their uses in industrial processes

6.1 Activation energy

In order to start a reaction, chemical bonds must be broken so that new bonds can be formed. The energy necessary to break these bonds is the **activation energy** of the reaction. Look at the graph in Figure 1.14. It shows the energy changes which take place when hydrogen peroxide breaks down to produce water and oxygen: $2H_2O_2 \rightarrow 2H_2O + O_2$

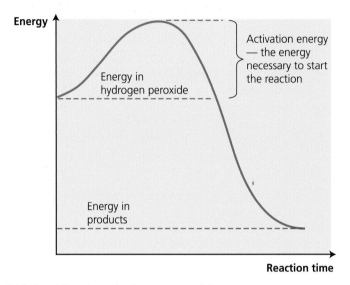

Figure 1.14 *Breaking down hydrogen peroxide*

One way in which we can provide the activation energy to start this reaction is by heating the hydrogen peroxide. Another way is to add an enzyme called catalase. This works in a different way. Catalase lowers the activation energy and, as a result, the reaction will take place at the much lower temperatures found inside the cells of living organisms.

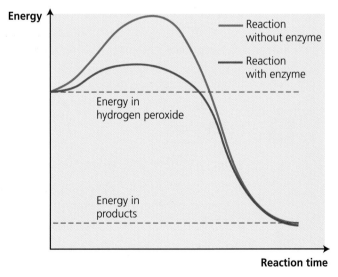

Figure 1.15 *The effect of an enzyme on activation energy*

Look carefully at Figure 1.15, in particular at the curve showing the energy changes when an enzyme is added. The curve starts and finishes at the same level; the only difference is that the activation energy is much less with an enzyme than without.

6.2 How enzymes work

The active site is part of the enzyme molecule, not part of the substrate.

Enzymes are proteins and, like any other proteins, are made up of a particular sequence of amino acids. This sequence results in a chain of amino acids which is folded into a particular tertiary structure. A small number of the amino acids, making up the enzyme molecule, form what is known as the **active site**. This is like a pocket in the surface of the enzyme into which a substrate molecule will fit and form an **enzyme–substrate complex**. The formation of this complex allows the reaction to take place. The products are released from the active site, leaving the enzyme free to combine with another substrate molecule.

This is the general idea. Biologists have developed models to explain what happens in a little more detail. Figure 1.16 shows two of these models.

Lock and key hypothesis

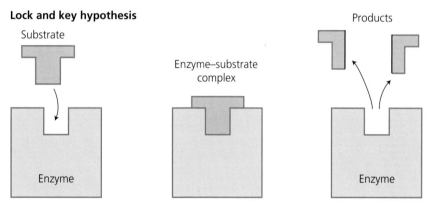

- The enzyme is a protein. The tertiary structure of the protein results in the active site of the enzyme having a specific shape.
- The substrate fits the active site and an enzyme–substrate complex is formed.
- The reaction takes place and the products are released.

Induced fit hypothesis

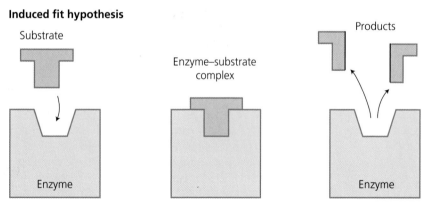

- The induced fit hypothesis is very similar to the lock and key hypothesis. The difference is that the active site moulds round the substrate rather like a sock on a foot.

Figure 1.16

6.3 Enzymes are specific

One of the consequences of the way in which enzymes function is that they are very specific as to the substrates on which they work. Different molecules have different shapes. Only the substrate of a particular enzyme will have the right shape to fit into the enzyme's active site.

6.4 Enzymes and organisms

Enzymes are extremely important because:

- They allow reactions to take place at the relatively low temperatures that are found in cells. Without enzymes, most biochemical reactions would take place very slowly.
- They control metabolic pathways. A metabolic pathway involves a sequence of reactions. Substance A, for example might be converted to either Substance D or to Substance F through a number of intermediate stages as shown in Figure 1.17. Each of these steps will be controlled by a different enzyme. We can call them enzyme b, enzyme c, and so on. By controlling the activity of different enzymes, the amounts of the various substances in the pathway can be regulated precisely.

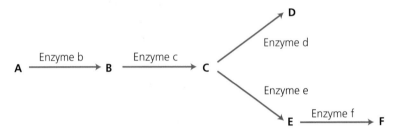

Figure 1.17

6.5 Making use of enzymes

Enzymes are widely used in industry and medicine. Table 1.4 shows some of the reasons why they are particularly useful.

Enzymes are specific	When starch is boiled with an acid, it is hydrolysed. A mixture of different products is formed, each made up of a different number of sugar units. However, when starch is hydrolysed by the enzyme maltase, only one product, maltose, is formed. The specificity of enzymes also makes them very useful in genetic engineering. There are many different restriction endonuclease enzymes. Each cuts the DNA in a particular place — so DNA can be cut in different ways with different enzymes.
Enzymes function at low temperatures	The Haber process is a chemical process which produces ammonia from nitrogen. It needs high temperatures and pressures. Nitrogen-fixing bacteria can convert nitrogen to ammonia at ordinary temperatures and pressures. These bacteria can do this because they rely on an enzyme, nitrogenase. Although we are at present unable to make use of nitrogenase in industrial processes, we can use other enzymes to save on fuel costs.
There are many different enzymes	Converting one organic compound into another may be extremely difficult if we rely only on chemistry. If many stages are involved, there may only be a very low yield at the end. There are many different enzymes, and one of these may be able to catalyse the reaction concerned directly.
Enzymes are proteins	At the end of any chemical reaction there are waste products. These may be corrosive or poisonous and difficult to get rid of. Enzymes are proteins and can be broken down by naturally occurring microorganisms.

Table 1.4 Properties and uses of enzymes

7 The properties of enzymes

When you have finished revising this topic, you should be able to explain how each of the following affects the rate of an enzyme-controlled reaction:

- temperature
- pH
- inhibitors
- substrate concentration

7.1 Active sites and enzyme action

When you answer questions about active sites and enzyme action, you need to make sure that you bring in the two important ideas — shape and fit.

An enzyme-controlled reaction depends on substrate molecules fitting into the active site of the enzyme. Environmental factors which alter the shape of the active site will therefore alter the rate of the reaction. These factors include high temperatures, variations in pH and the presence of inhibitors.

High temperature

The effect of high temperature on an enzyme-controlled reaction can be summarised as follows:

Heating breaks the bonds that maintain the enzyme's tertiary structure. It does not break the peptide bonds.

- High temperatures result in an increase in the kinetic energy of the enzyme molecules. They vibrate, and the chemical bonds that maintain the tertiary structure of the enzyme molecule break.
- This results in the enzyme molecule changing shape; in other words, the enzyme is **denatured**.
- Therefore the shape of the active site alters.

Use the word 'denatured'. Enzymes are never killed by heating!

- The substrate molecule does not fit into the active site, so an enzyme–substrate complex is not formed.

pH

Changing the pH outside the limited range in which the enzyme normally works has similar consequences. The only real difference is that a change in pH alters the number of free hydrogen (H^+) or hydroxyl (OH^-) ions. This affects the charge on the amino acid residues that make up the enzyme. It is this that leads to denaturation and the change in the shape of the active site of the enzyme. Figure 1.18 shows the affect of pH on a typical enzyme from a human cell.

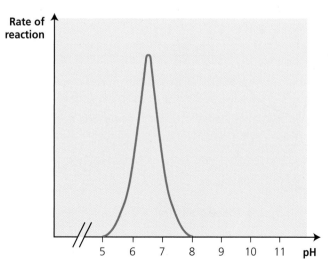

Figure 1.18 The effect of pH on the rate of an enzyme-controlled reaction

Inhibitors

Inhibitors slow down the rate of enzyme-controlled reactions. Their action is described as either **competitive** or **non-competitive**. This is summarised in Figure 1.19.

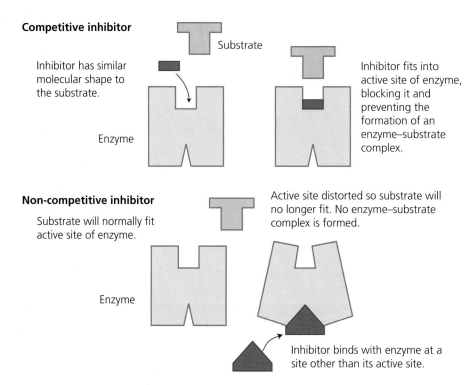

Figure 1.19 *The action of competitive and non-competitive inhibitors*

Inhibitors are very important in controlling metabolic processes. The amount of a particular substance produced by a metabolic pathway has to be matched closely to the needs of the organism. If too much or too little is produced, some form of metabolic disorder will result. In addition, producing more of a substance than is required wastes resources. In many metabolic pathways, the product can inhibit one of the enzymes at the beginning of the pathway, so that the pathway becomes self-limiting. This is summarised in Figure 1.20.

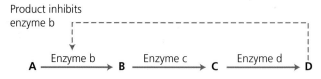

- If a large amount of substance D has been produced, enzyme b will be inhibited. Less substance D is produced.

- If there is too little substance D, enzyme b will not be inhibited. More substance D will be produced.

Figure 1.20 *Controlling metabolic pathways*

7.2 Collisions

Before an enzyme-controlled reaction can take place, the enzyme and its substrate must come together. They must collide with each other; the more

collisions in a given period of time, the faster the rate of reaction. The chance of a successful collision taking place is greatly increased by increasing the temperature and the concentration of the substrate.

Temperature

We have already seen that high temperatures denature enzymes and stop them from working. At temperatures below those that bring about denaturing, an increase has a different effect. A rise in temperature from, say, 20°C to 30°C brings about an increase in the rate of reaction as follows:

- Increasing the temperature increases the kinetic energy of the molecules. They move around faster.
- This increases the chance of substrate and enzyme molecules colliding with each other.
- The more collisions in a given period of time, the faster the rate of reaction.

In most enzyme-controlled reactions, the rate of reaction more or less doubles with a 10°C rise in temperature. This relationship can be described mathematically by calculating the **temperature coefficient** or Q_{10}.

Calculating Q_{10}

Catalase breaks down hydrogen peroxide to produce water and oxygen. A mixture of catalase and hydrogen peroxide gave off 24 cm^3 of oxygen in 1 minute at 20°C. An identical mixture gave off 50 cm^3 of oxygen in 1 minute at 30°C.

$$Q_{10} = \frac{\text{rate at } 30°C}{\text{rate at } 20°C} = \frac{50}{24} = 2.1$$

Most enzyme-controlled reactions have a Q_{10} of approximately 2.

An increase in temperature therefore increases the rate at which the enzyme works until it reaches its optimum value. Above this, an increase in temperature results in the enzyme being denatured and the rate of reaction decreasing. This is summarised in Figure 1.21.

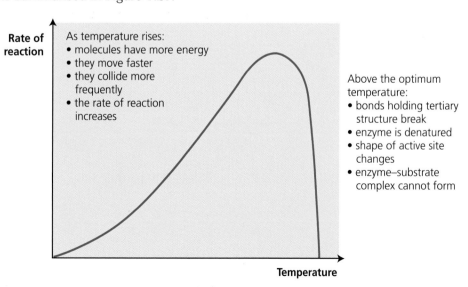

Rate of reaction

As temperature rises:
- molecules have more energy
- they move faster
- they collide more frequently
- the rate of reaction increases

Above the optimum temperature:
- bonds holding tertiary structure break
- enzyme is denatured
- shape of active site changes
- enzyme–substrate complex cannot form

Temperature

Figure 1.21 The effect of temperature on the rate of an enzyme-controlled reaction

Substrate concentration

Figure 1.22 shows what happens to the rate of a reaction when the concentration of the substrate is increased.

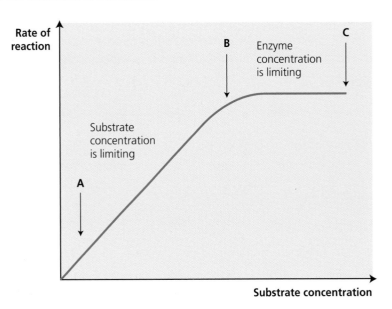

Figure 1.22 The effect of substrate concentration on the rate of an enzyme-controlled reaction

Look first at the part of the curve between **A** and **B**. It shows us that as the concentration of substrate increases, so does the rate of the reaction. In other words, the substrate concentration is limiting the rate of the reaction. The more substrate there is, the greater the chance of a substrate molecule colliding with an enzyme and a reaction taking place.

Now look at the part of the curve between **B** and **C**. Increasing the substrate concentration makes no difference to the rate of reaction. At any one time, all the enzyme active sites are taken up. The enzyme cannot work any faster, and the only way that the rate of reaction can be increased is to increase the number of enzyme molecules. Between **B** and **C**, the enzyme concentration is said to be limiting.

CHAPTER 2 Cells

1 Investigating cell structure

When you have finished revising this topic, you should be able to:

- describe the advantages and limitations of using optical and electron microscopes in studying the structure of cells
- list the main differences between prokaryotic and eukaryotic cells
- calculate the magnification or actual size of a structure from a drawing or a photograph
- describe and explain the importance of the steps involved in using a centrifuge to separate cell organelles
- list the main differences between the structures of plant and animal cells

1.1 Different types of microscope

'Magnification' refers to greater size; 'resolution' refers to greater detail. Make sure that you can distinguish between these terms.

The most powerful **light microscopes** magnify up to 1500 times. They allow individual cells and some of the structures within them, such as nuclei, to be seen. The problem with a light microscope is that it has limited **resolution**. The resolution of a microscope is its ability to distinguish between objects that are close together. There is no point in having magnification without resolution. If an object looks small and rather fuzzy with a microscope, increasing the magnification will make it look larger but just as fuzzy. Despite this, light microscopes do have advantages in letting us examine living cells and tissues.

Electron microscopes have much greater resolution so they allow us to see the detailed structure of cell organelles such as mitochondria and chloroplasts. In order to get increased resolution, a beam of electrons is used instead of light as it has a smaller wavelength. Electrons are very small and would scatter if they hit molecules present in the air. Because of this they must travel through a vacuum. Specimens have to be cut very thin and treated in particular ways so that they can be examined under these conditions. This treatment means that the appearance of the specimen may be changed during preparation. Changes in appearance produce **artefacts** and care needs to be taken to distinguish the actual specimen from an artefact produced during its preparation.

Feature	Light microscope	Electron microscope
Method of illumination	Light	Beam of electrons
Method of focusing	Glass lenses	Magnets
Specimen	May be alive or dead	Will be dead
Magnification	Maximum magnification of a research microscope is about 1500 times; most student microscopes have a maximum magnification of about 400 times.	Up to 500 000 times
Resolution	About 0.2 μm	About 0.001 μm or 1 nm

Table 2.1 The main features of light and electron microscopes

There are two main types of electron microscope. In a **transmission electron microscope**, a beam of electrons is passed through an object and shows the internal structure in detail. In a **scanning electron microscope**, the electron beam is reflected off the surface. It produces a three-dimensional view of the surface. Table 2.1 summarises some of the main features of light and electron microscopes.

1.2 Prokaryotic and eukaryotic cells

Microscopes have allowed us to recognise many different types of cell, but there are major differences between the **prokaryotic** cells which are found in organisms like bacteria, and the **eukaryotic** cells found in all other organisms. Table 2.2 summarises some of the more important differences between these types of cells.

Feature	Prokaryotic cells	Eukaryotic cells
Size	Cells smaller, usually less than 5 µm in diameter.	Cells larger, often as much as 50 µm in diameter.
Nucleus and DNA	Cells do not have a nucleus. The DNA, which is not associated with proteins, is present as a circular strand in the cytoplasm of the cell.	Cells have a nucleus. The DNA, which is in long strands, is associated with proteins forming chromosomes.
Organelles	Few organelles present and none of them are surrounded by a plasma membrane.	Many membrane-surrounded organelles, such as mitochondria, present.
Ribosomes	Only have small ribosomes which are free in the cytoplasm.	Have small ribosomes and larger ones which are associated with membranes forming rough endoplasmic reticulum.

Table 2.2 Differences between prokaryotic and eukaryotic cells

1.3 Calculating the actual size of a cell

If we know the magnification, we can calculate the actual size of a cell from a drawing or a photograph. Look at the example below.

In calculating the actual size from the observed size, measure the observed size in millimetres. It is easier to remember:

1 km = 1000 m

1 m = 1000 mm

1 mm = 1000 µm

Example

The cell in Figure 2.1 has been magnified 1800 times. Calculate its actual size. Give your answer in micro-metres (µm).

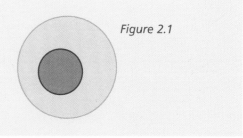

Figure 2.1

Set out calculations clearly. If you make an arithmetical mistake, the examiners may still be able to give credit.

$$\text{magnification} = \frac{\text{observed size}}{\text{natural size}}$$

Therefore

$$\text{natural size} = \frac{\text{observed size}}{\text{magnification}} = \frac{27\,\text{mm}}{1800}$$

But we want the answer in micrometres. There are 1000 micrometres in 1 millimetre, so

$$\text{natural size} = \frac{27000\ \mu\text{m}}{1800} = 15\ \mu\text{m}$$

Obviously, if you need to calculate magnification, all you need to do is to go back to the original formula and divide the observed size by the real or natural size.

1.4 Cell fractionation

Cell organelles can be separated from each other in the process of **cell fractionation**. In outline, this involves breaking up a suitable sample of tissue and centrifuging it at different speeds. The flow chart summarises the key steps in the procedure.

The tissue is broken up in a **homogeniser**, a machine rather like a kitchen blender. It is suspended in a buffer solution which keeps the pH constant. This solution is kept cold and has the same water potential as the tissue.

↓

The mixture obtained is filtered. This removes any large pieces of tissue that were not broken up in the homogeniser.

↓

The filtrate is now centrifuged at low speed. Larger organelles such as nuclei and chloroplasts fall to the bottom of the centrifuge tube where they form a pellet. They can be resuspended in a fresh solution if they are required.

↓

The liquid or supernatant can be spun in the centrifuge again, at a higher speed. This results in smaller organelles such as the mitochondria separating out into a pellet.

The **buffer solution** is added to the tissue to keep the pH constant. Changes in pH affect enzymes and other proteins which are required for the functioning of the organelles which are being isolated. The solution is kept cold so that enzymes are not denatured. The aim of using a solution with the same water potential as the tissue is to prevent damage to the organelles caused by water entering or leaving them by osmosis.

1.5 Plant and animal cells

Plants and animals are both eukaryotes. Their cells are similar in many ways but there are also some important differences. Table 2.3 summarises the main features of plant and animal cells. You must bear in mind when looking at this table, however, that there is no such thing as a 'typical' cell. A red blood cell, for example, differs from most other animal cells as it has no nucleus and no mitochondria. Mesophyll cells from a plant leaf contain large numbers of chloroplasts but there are no chloroplasts in root cells.

Be careful to distinguish between cell membranes and cell walls. Cell walls are never found in animal cells

Feature	Present in		Main function
	plant cell	animal cell	
Cell wall	✔		Cell walls provide strength. They resist the pressure produced from the entry of water by osmosis.
Plasma membrane	✔	✔	Controls the passage of substances into and out of the cell.
Nucleus	✔	✔	Contains the DNA which holds the genetic information necessary for controlling the cell.
Mitochondrion	✔	✔	Produces ATP from aerobic respiration.
Chloroplast	✔		Photosynthesis.
Ribosome	✔	✔	Protein synthesis.
Rough endoplasmic reticulum	✔	✔	Collection and transport of proteins.
Smooth endoplasmic reticulum	✔	✔	Synthesis of lipids.
Golgi apparatus	✔	✔	Packaging and processing of molecules such as proteins synthesised in the cell. Formation of lysosomes. Secretion of cell wall carbohydrates.
Lysosome	✔	✔	Digestion of unwanted material in the cell.

Table 2.3 Main features of plant and animal cells

1.6 Cell specialisation

Most eukaryotic cells in multi-cellular organisms are adapted to do a particular job. For example, epithelium cells in the small intestine are adapted to absorb nutrients efficiently and palisade mesophyll cells in leaves show a range of adaptations for efficient photosynthesis.

Similar cells which are specialised for a particular function are organised into **tissues**. These include epithelium, muscle and connective tissue in animals, and xylem and phloem in plants. Tissues are organised into **organs**. An organ is defined as a group of tissues that work together to perform a particular function. For example, a leaf is an example of a plant organ. Organs themselves may be organised into an **organ system.** For example, the stomach, small intestine and large intestine are organs that make up the digestive system.

2 *Cell membranes and their structure*

When you have finished revising this topic, you should be able to describe:

■ the fluid-mosaic model showing the structure of a cell membrane

■ the arrangement of phospholipid molecules in a membrane and explain the importance of this arrangement

■ the arrangement of proteins in the membrane and list some of their functions

2.1 Introduction

Cells exchange a variety of substances with their surroundings. These substances pass into or out of the cell through the **plasma membrane**. This membrane surrounds all cells and is very similar in structure to the membranes around organelles, such as the nuclei and the mitochondria in eukaryotic cells. Substances pass through the membrane by a variety of processes; so in order to understand these, we need to look first at its structure.

2.2 The fluid-mosaic model

If asked to explain why a cell membrane may be described as a fluid-mosaic, make sure that you explain both why it is fluid **and** why it is a mosaic.

The plasma membrane is only about 7 nm thick. It is not possible to see its detailed structure, even with an electron microscope, so biologists have produced a model to explain its properties. This is known as the fluid-mosaic model. It has been given this name because it suggests that the positions of the molecules within it are fluid. They are able to move around within the membrane. The model also suggests that the membrane is made up from a variety of different substances arranged in a mosaic.

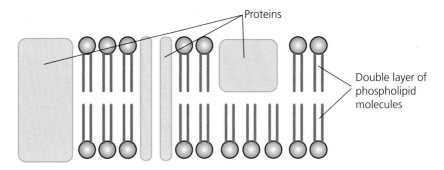

Figure 2.2 The structure of the plasma membrane

Figure 2.2 is a diagram showing a simplified version of the structure. It is based on a **phospholipid bilayer**. This is a double layer of phospholipid molecules arranged with their hydrophobic tails on the inside of the membrane. Molecules that dissolve in lipids can pass easily through this bilayer. Water soluble molecules, however, must pass through pores in the protein molecules that span the membrane. The phospholipid layer forms, therefore, a very important barrier. Since molecules of many substances are unable to pass through it directly, passage into and out of the cell can be effectively regulated by the protein molecules in the membrane.

Figure 2.2 also shows the presence of these **protein** molecules. Some of these molecules float freely in the phospholipid bilayer, while others are attached more firmly to structures in the cytoplasm of the cell. Membrane proteins have a variety of different functions:

- They may act as enzymes. Some of the enzymes that digest carbohydrates are found on the plasma membrane of the cells that line the small intestine.
- They act as carrier proteins and play an important part in the processes of facilitated diffusion and active transport.
- They act as receptors for hormones. A hormone will only act on a cell if it has the right protein receptors in its plasma membrane.

Other substances are also present in the membrane. They include **cholesterol**, which is found between the phospholipid molecules. Plasma membranes containing a relatively large amount of cholesterol are not very permeable to water or to small inorganic ions such as those of sodium. Carbohydrates are attached to lipids and proteins on the outside surface of the membrane forming **glycolipids** and **glycoproteins**. These are important in allowing cells to recognise one another.

All membranes have this same basic structure but the proportion of different molecules differs slightly from cell to cell. This difference in composition is related to function.

3 The cell cycle

When you have finished revising this topic, you should be able to:

- explain how a molecule of DNA replicates to form two identical molecules
- explain the importance of mitosis
- describe the behaviour of chromosomes during mitosis
- describe the events of the cell cycle

3.1 Replicating DNA

The DNA inside the single cell, (formed when a sperm fuses with an egg) contains the genes necessary to code for all the proteins produced by an organism in its lifetime. These genes can be switched on and off according to need. It is therefore important that every cell in an organism has an exact copy of this DNA. Figure 2.3 shows how **replication** of DNA takes place.

> Make sure that you distinguish between DNA replication in which more molecules of DNA are made, and transcription in which a molecule of mRNA is produced.

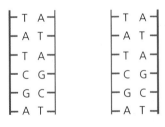

- The DNA molecule unwinds at many places along its length.
- The hydrogen bonds holding complementary bases together break, and the polynucleotide strands separate. This process involves a number of enzymes and other proteins.

- Each of these DNA strands acts as a template for the formation of a new molecule of DNA.
- Individual nucleotides found in the nucleus line up with the complementary bases on the parent DNA strands.

- The nucleotides are joined together by an enzyme called DNA polymerase to form two molecules of DNA. Each molecule contains one of the polynucleotide strands from the parent molecule. The other strand is a new one. This method of replication is known as **semi-conservative replication**.

Figure 2.3 The replication of DNA

The replication of DNA takes place before the chromosomes divide during mitosis.

3.2 Mitosis

Mitosis is the form of nuclear division that takes place during the growth and development of an organism. If we think of the chromosomes as a way of packaging the DNA present in a cell, then mitosis ensures that the chromosomes and the DNA they contain are split equally when the cell divides.

Consider a simple cell in which the diploid number of chromosomes is four. There are two pairs of chromosomes. One of the chromosomes of each pair came originally from the male parent, while the other came from the female parent. Nearly all cells have more than four chromosomes, but using a small number makes the process easier to follow. Figures 2.4a–e - shows mitosis in this cell.

Don't make the mistake of thinking that all organisms are like humans in having 23 pairs of chromosomes in each body cell.

You will need to know the order of the different stages of mitosis. The initial letters make IPMAT — it doesn't mean anything but it is easy to remember.

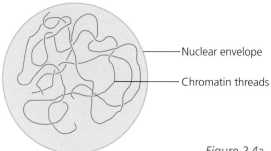

Figure 2.4a Interphase

This is the stage between mitotic divisions. Although individual chromosomes are not visible in the nucleus, a number of events are taking place in the cell that is essential for mitosis. These include:

● The replication of DNA.
● The production of ATP, the separation of chromosomes during mitosis requires energy.
● The synthesis of proteins required for cell division.

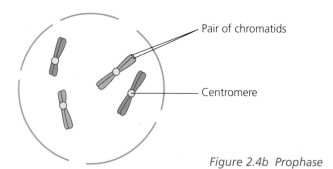

Figure 2.4b Prophase

● The chromosomes start to become visible. Each consists of a double structure made up of a pair of identical **chromatids**. They are joined at the **centromere**.
● The nuclear envelope starts to break down.

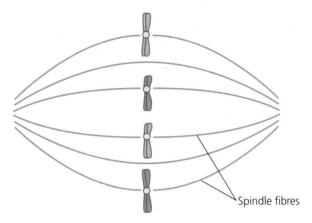

Figure 2.4c Metaphase

- The chromosomes are arranged across the middle or the equator of the cell and are attached to some of the protein fibres that form the **spindle**.
- The nuclear envelope has now disappeared.

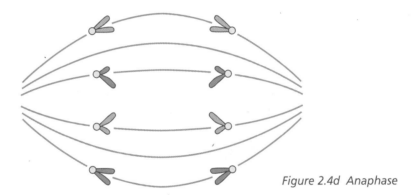

Figure 2.4d Anaphase

- The chromatids are moving apart, pulled by the spindle fibres to the opposite poles of the cell. One chromatid from each pair goes to each pole. Once the chromatids have separated from each other, they are called chromosomes.

Figure 2.4e Telophase

- The two groups of chromosomes come together at opposite ends of the cell, each group containing a complete set of chromosomes, identical to those of the parent cell.
- The chromosomes now start to uncoil and lose their distinct appearance.
- New nuclear envelopes form.

3.3 The cell cycle

It is possible when looking at a series of diagrams such as those in Figure 2.4 to lose sight of the fact that mitosis is a continuous process. The stages of mitosis run into each other and form part of the cycle of events in which cells grow and divide. This is the **cell cycle**, and it is summarised in Figure 2.5.

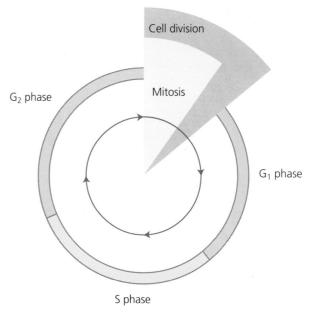

Figure 2.5 The cell cycle

There are four main stages in the cell cycle. Three of these stages, or phases, take place during the interphase:

1 **The first growth phase, G_1.** The cells which form as a result of mitosis are small. In the period immediately after they have divided, they start to grow. During this phase, protein synthesis takes place and the cell organelles increase in number. Cells which do not divide, such as nerve cells in the brain, are nearly always in this phase.

2 **The DNA synthesis phase, S.** This is the stage of the cell cycle in which DNA replication takes place.

3 **The second growth phase, G_2.** The second growth stage takes place immediately before mitosis. In this phase, the proteins required for cell division are synthesised.

4 **Mitosis and cell division.**

3.4 Distinguishing mitosis and meiosis

The key differences between mitosis and meiosis (which is the type of cell division involved in the production of haploid gametes) are summarised in Chapter 6. The important difference to remember at this point, however, is that daughter cells formed during mitosis have identical copies of genes while cells produced as a result of meiosis are not genetically identical.

DNA replication takes place during interphase, not during mitosis.

CHAPTER 3

Exchange and transport

Diffusion, osmosis and active transport

When you have finished revising this topic, you should:

- be able to explain what is meant by diffusion and describe the differences between simple diffusion and facilitated diffusion
- understand how various factors affect the rate of diffusion
- understand the meaning of the term 'water potential' and be able to define osmosis in terms of the movement of water from a high water potential to a lower one
- understand what is meant by active transport and be able to explain its importance in transporting substances into and out of cells

1.1 Diffusion

The molecules present in liquids and gases are always moving. Because this movement is at random, more molecules will move from where they are in high concentration to where they are in a lower concentration than the other way round. This is **diffusion**, the net movement of molecules from a high concentration to a low concentration. The movement of the molecules relies on kinetic energy. They don't need an input of energy from the cell in the form of ATP. Because of this, diffusion is said to be a **passive** process.

> Diffusion does not require the presence of a membrane.

Diffusion plays a vital role in the movement of substances in living organisms. It is the main way in which oxygen, for example, moves from the alveoli of the lungs into the blood and from capillaries into respiring cells. It is, however, an extremely slow process, so it is useful only when the distances involved are very small.

Small molecules, such as those of oxygen and carbon dioxide, are able to diffuse through the plasma membrane into or out of a cell. It seems that they may be able to pass between phospholipid molecules in the bilayer, and this is helped by the phospholipids continually moving around.

1.2 Facilitated diffusion

Molecules such as those of glucose are unable to pass directly through the phospholipid bilayer. They rely on the presence of carrier proteins present in the membrane. These **carrier proteins** have two important properties:

1 They have a **binding site**. This binding site is specific. A glucose carrier, for example, will have a binding site into which only glucose molecules will fit. In addition, different sorts of cells have different sorts of carriers. This explains why a particular cell will take up some substances but not others.

2 They are able to change shape. In one form the binding site is exposed to the outside; in the other it is on the inside. So, taking the example of a glucose molecule, it will fit into a carrier protein which has its binding site on the outside of the cell; the carrier protein changes shape and the glucose molecule is released on the inside of the cell. This is summarised in Figure 3.1.

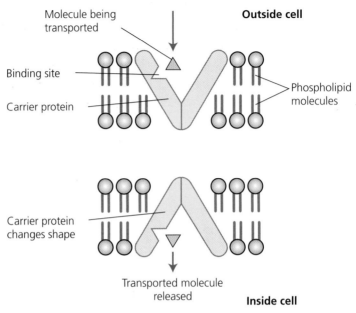

Figure 3.1 Facilitated diffusion

As with simple diffusion, facilitated diffusion relies on the **kinetic energy** of the molecules. It is also a passive process and does not need an input of energy from the cell in the form of ATP.

1.3 The rate of diffusion

Various factors affect the rate of diffusion. These can be summarised by **Fick's law**. In its simplest form this says that:

$$\text{the rate of diffusion is proportional to} = \frac{\text{surface area} \times \text{difference in concentration}}{\text{thickness of the exchange surface}}$$

A surface that is adapted for efficient diffusion will:
- have a large area
- maintain a large difference in concentration on either side of the exchange surface
- have as thin an exchange surface as possible

This relationship is very important and you will come across it wherever you encounter diffusion, such as in the exchange of respiratory gases, and in the absorption of the products of digestion in the small intestine.

In addition to these three factors, temperature also affects the rate of diffusion. An increase in temperature increases the kinetic energy of the molecules. The higher the temperature, the faster the molecules will be moving, so the greater the rate of diffusion.

Most exchange surfaces over which diffusion takes place are moist. The reason for this is simple. The surface has to be permeable otherwise substances could not diffuse across it. If it is permeable to small molecules such as oxygen and carbon dioxide, it will be permeable to water. Water, then, will diffuse out of the cells — so the exchange surface is bound to be moist.

Whenever you come across diffusion, think about Fick's law. What are the adaptations that ensure a large surface area, a large difference in concentration and a thin exchange surface?

Remember, exchange surfaces are moist **so that** diffusion can take place.

1.4 Water potential, ψ

Look at Figure 3.2. It shows water molecules surrounded by a membrane. These water molecules are in constant motion. As they move about randomly some of them will hit the membrane. The collision of the molecules with the membrane will create a pressure on it. This pressure is known as the **water potential** and it is measured in units of pressure: kilopascals (kPa) or megapascals (MPa).

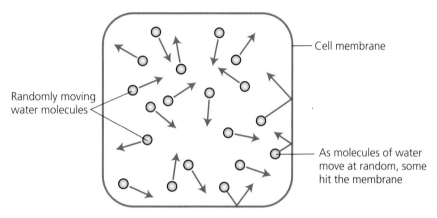

Figure 3.2

It follows from this that the more water molecules that are present and able to move about freely, the greater the water potential. The greatest number of water molecules that it is possible to have is in distilled water — because this is pure water and there is nothing else present. Distilled water therefore has the highest water potential. This is given the value of zero. All other solutions will have a value less than this. They will have a **negative water potential**.

Now look at Figure 3.3. This shows the situation in a cell that is surrounded by water. Separating the cytoplasm of the cell from the surrounding solution is the plasma membrane. It is **partially permeable**. This means that it allows small molecules such as water to pass through, but not larger molecules. The cytoplasm contains many soluble molecules and ions. They attract water molecules to form a 'shell' around them. These water molecules can no longer move around freely in the cytoplasm, so there is a much higher concentration of free water molecules in the surrounding water than there is in the cytoplasm. The water potential of the water surrounding the cell is higher than that of the cytoplasm. Water molecules will diffuse from a solution with a higher water potential to a solution with a lower water potential. This is **osmosis**. We can, therefore define osmosis in terms of water potential:

Osmosis is the net movement of water molecules from a solution with a higher water potential to a solution with a lower water potential through a partially permeable membrane.

> A solution with a lower water potential is a solution with a more negative water potential.

> Osmosis involves the movement of **water**.

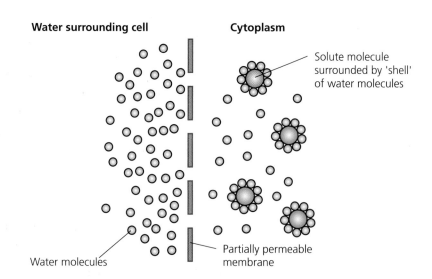

Figure 3.3

Hyper means more; hypo means less ... and hyper has more letters than hypo.

Suppose two solutions have different concentrations of dissolved substances present. The one with the higher concentration is described as being **hypertonic** to the one with the lower concentration, while that with the lower concentration is described as being **hypotonic**.

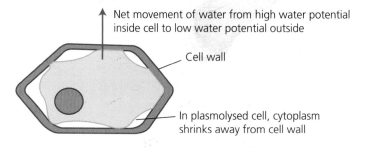

Figure 3.4

Figure 3.4 shows a plant cell in a hypertonic sucrose solution. The water potential of the cell is higher (less negative) than that of the surrounding solution, so water will move out of the cell by osmosis. You should note that a plant cell is surrounded by a rigid cell wall made of cellulose. As water moves out of the cell, its cytoplasm will shrink away from the cell wall. The cell is said to be **plasmolysed**.

If, on the other hand, this cell is placed in a hypotonic solution, water will move into the cell. Its cytoplasm will swell and push against the surrounding cell wall. The cell is now described as being **turgid**. The fact that plant cells can take in water by osmosis and become turgid is extremely important in providing support to plants.

A cell wall is not a special kind of membrane found in plant cells. Both plant cells and animal cells have plasma membranes; only plant cells have a cell wall.

Animal cells do not have this cell wall. So, if a cell such as a red blood cell is placed in distilled water, water will move in, and it will increase in size and burst. Obviously, it is very important that the water potential of animal tissues is kept constant and most animals have ways of doing this.

1.5 Active transport

Most cells need substances that are only present in low concentrations in their environment. Plants, for example need mineral ions which are only present in small amounts in the surrounding soil. **Active transport** is a process that enables a cell to take up a substance against a concentration gradient. Like facilitated diffusion, it requires the presence of protein carrier molecules which transport the substance across the membrane. It also requires energy; this comes from the ATP produced during respiration. Cells, where a lot of active transport takes place, such as those which line the small intestine, have large numbers of mitochondria which produce the necessary ATP.

1.6 Comparing diffusion and active transport

The similarities and differences between the processes which we have been looking at are summarised in Table 3.1.

	Simple diffusion	Osmosis	Facilitated diffusion	Active transport
Require protein carrier molecules			✔	✔
Passive process not requiring ATP	✔	✔	✔	
Able to transport substance from a low concentration to a high concentration				✔
Process stopped by a respiratory poison				✔

Table 3.1 Comparison of diffusion and active transport

2 *Gas exchange*

When you have finished revising this topic, you should:

- be able to describe the relationship between the size of an organism and its surface area to volume ratio, and explain the importance of this in gas exchange

- understand that all organisms rely on diffusion for the exchange of respiratory gases

- be able to explain how the efficiency of gas exchange is increased by:
  - a large surface area
  - a large difference in concentration across the gas exchange surface
  - a thin gas exchange surface

- understand that gas exchange in terrestrial organisms always involves a loss of water from the exchange surface

- be able to describe the main features of the gas exchange surface in a mammal and in a flowering plant

2.1 Size and surface area to volume ratio

Table 3.2 shows some features of cubes of different size. For simplicity, the units have been omitted.

Side length	Total surface area	Volume	Surface area / Volume
1	6	1	6
2	24	8	3
3	54	27	2

Table 3.2

> This is a very important principle. The larger an organism, the smaller its surface area to volume ratio. Make sure you get it the right way round.

Obviously, the bigger the cube, the larger its surface area; an elephant will clearly have a larger surface area than a mouse. If you look at the last column of the table, however, you will see that the bigger the cube, the smaller is its surface area to volume ratio. An elephant has a smaller surface area to volume ratio than a mouse.

Diffusion depends on surface area, so a very small organism such as an amoeba has a large surface area to volume ratio and can meet all its exchange requirements by diffusion through its surface. Larger organisms such as insects, fish and mammals have relatively small surface area to volume ratios. They have specialised gas exchange systems (and other systems such as the gut) that increase the surface area over which diffusion can take place.

2.2 The factors determining the rate of diffusion

There are various factors that affect the rate of diffusion through a gas exchange surface. These are summarised by **Fick's law** (see page 46):

$$\text{rate of diffusion} \propto \frac{\text{surface area} \times \text{difference in concentration on either side of surface}}{\text{thickness of gas exchange surface}}$$

The most efficient gas exchange surface will be the one over which the rate of diffusion will be as high as possible, so it will have adaptations which:

- provide a large surface area
- maintain a large difference in concentration on either side of the surface
- ensure that it is as thin as possible

You may not know this relationship as Fick's law, but the principle is very important and it will provide you with a useful framework for looking at the important features of any gas exchange surface.

> Gas exchange surfaces are moist because water molecules will diffuse through them; they are not moist in order to increase the rate of diffusion.

There is one additional feature that we should consider here. Gas exchange surfaces in terrestrial organisms are always moist. There is a simple reason for this. For diffusion to take place, the surface must be permeable and allow small molecules such as those of oxygen and carbon dioxide to pass through. Water molecules are also small, and they are at a much greater concentration inside the organism than outside it. This means that if oxygen can diffuse in, water can diffuse out, and the exchange surface will be moist.

If you have to study gas exchange in other organisms, use the same framework. Explain how the gas exchange organs provide a large surface area, maintain a difference in concentration and have a thin exchange surface.

2.3 Gas exchange in a mammal

The lungs of a mammal form a very efficient gas exchange surface because they are specially adapted.

They provide a large surface area
- The very large number of tiny air sacs or **alveoli** provide a huge surface area.
- The lungs receive as much blood in a minute as all the rest of the body put together. This blood flows through a capillary network; the total surface area of all these lung capillaries is also very large.

They maintain a large difference in concentration on either side of the exchange surface
- A ventilation mechanism ensures that fresh air is taken in with each breath. This means that the oxygen and carbon dioxide diffusion gradients across the exchange surface will remain high.
- The circulation of blood through the pulmonary circulation ensures that blood saturated with oxygen and low in carbon dioxide is replaced with blood low in oxygen and high in carbon dioxide. This also helps to maintain the diffusion gradients across the exchange surface.

They have a thin exchange surface
- There are only two layers of cells between the air in the alveolus and the blood in the capillaries. These are the alveolar epithelium and the cells forming the capillary wall.
- The cells that form the alveolar lining are flat and very thin. Cells that are this shape are known as **squamous epithelial cells**. Respiratory gases have only to diffuse a distance of about 0.2 μm to pass through these cells.

2.4 Gas exchange in a flowering plant

The key structures within the leaf that allow the exchange of oxygen and carbon dioxide are known as **stomata**. These pores are found in the plant's epidermis. Stomata are present in large numbers, especially on the underside of leaves. They are concerned with gas exchange for **respiration** and **photosynthesis**, and also with the loss of water vapour by **transpiration**.

Stomata can be opened or closed by guard cells according to the requirements of the plant, often being open in the day (to provide carbon dioxide for photosynthesis) and closed at night (to minimise water loss). The mechanism for stomatal opening can be summarised as follows:
- Daylight or low carbon dioxide concentrations stimulate the active transport of potassium ions into the guard cells.
- Starch in the chloroplasts of guard cells are converted to malate ions.
- Increased potassium ion and malate ion concentration in the guard cells reduced the **water potential** of these cells.
- Water is drawn in by **osmosis**, the guard cells become turgid and the stomata open.

The opposite happens in the dark or when carbon dioxide concentrations are high.

3   *The transport of respiratory gases*

When you have finished revising this topic, you should:

- understand that large animals require a blood system to transport respiratory gases
- be able to describe the role of haemoglobin in transporting oxygen from the lungs of a mammal to its respiring tissues
- be able to explain what is meant by the Bohr shift and describe its importance in increasing the oxygen made available to respiring tissues
- be able to relate the oxygen-carrying properties of different types of haemoglobin to the environment in which the animal lives
- be able to describe how carbon dioxide is transported in the blood

3.1 Body size and blood systems

Oxygen passes through gas exchange surfaces and enters respiring cells by diffusion. Diffusion is an efficient means of transport if the distances involved are very small. If distances are large, however, it is far too slow. Large organisms have blood systems which aid the process of diffusion in ensuring that all the cells present in the organism receive an adequate supply of oxygen. This is summarised in Figure 3.5.

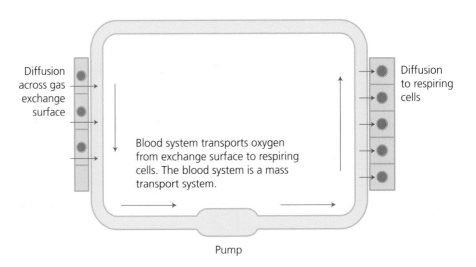

Figure 3.5 Supplying the cells of a large organism with oxygen involves diffusion and a blood system

3.2 The oxygen dissociation curve

Look at Figure 3.6. It shows a dissociation curve for human haemoglobin.

The *y*-axis shows the percentage saturation of haemoglobin with oxygen. It ranges from 0% when no oxygen is being carried to 100% when the haemoglobin is carrying all the oxygen it can. At 100% all the haemoglobin is present as oxyhaemoglobin. The *x*-axis shows the **partial pressure** of oxygen. In simple terms, this is a measure of how much oxygen is present.

When you are asked to explain how oxygen is transported by haemoglobin, make sure you mention the oxygen dissociation curve. Remember, only A-level detail will get A-level marks.

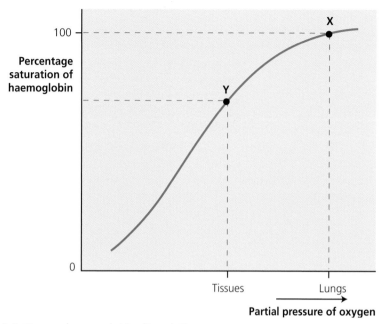

Figure 3.6 Human haemoglobin dissociation curve

Now look at the curve. When there is no oxygen present, none of the haemoglobin will be carrying oxygen. When the partial pressure of oxygen is at its highest, the haemoglobin will be saturated with oxygen. In between these points the curve is S-shaped.

So how does haemoglobin take up and release oxygen? The partial pressure of oxygen is higher in the lungs than it is in the tissues of the body. This is because air which is rich in oxygen enters the lungs and this oxygen is used in respiration. At the partial pressure of oxygen in the lungs, the curve shows us that haemoglobin is almost completely saturated with oxygen (point **X** on the curve). In other words, as blood flows through the lung capillaries, the haemoglobin is rapidly converted to oxyhaemoglobin.

In the tissues there is far less oxygen because it is being used up in respiration. Looking at the curve again, we can see that the percentage saturation of haemoglobin will be lower, so oxygen can be given up to the tissues (point **Y** on the curve).

3.3 The Bohr shift

Now we need to go a little further. In the body, the ability of haemoglobin to transport oxygen is also affected by the amount of carbon dioxide present (Figure 3.7).

Figure 3.6 shows an oxygen dissociation curve in which the partial pressure of carbon dioxide is low. In Figure 3.7 this curve has been drawn again and a second curve added to represent what happens when there is a high partial pressure of carbon dioxide. The more carbon dioxide there is, the more the oxygen dissociation curve is moved to the right. This is known as the **Bohr shift** and it has a considerable effect on the transport of oxygen.

Red blood cells do not change shape as they take up oxygen. The actual amount of oxygen transported by a red blood cell is very small and will not make any difference to the size or shape of the cell.

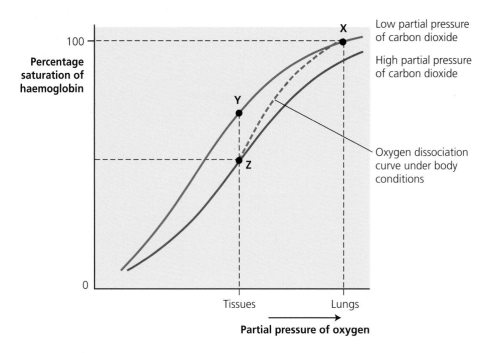

Figure 3.7 Carbon dioxide and the Bohr shift

The situation in the body then, is that in the lungs there is a high partial pressure of oxygen and a low partial pressure of carbon dioxide. The partial pressure of carbon dioxide will be low because this is where it is being removed from the body. We are still at point **X** on the graph and here, as we saw earlier, the haemoglobin will be saturated with oxygen.

In the tissues, there is a low partial pressure of oxygen and a high partial pressure of carbon dioxide. The partial pressure of carbon dioxide is high because it is being produced as a result of respiration. This means that we are on the lower curve at point **Z**, not on the upper curve at point **Y**. Consequently, the percentage saturation of haemoglobin will be lower and even more oxygen will be released to the tissues. The dotted line on the graph joins points **X** and **Z** and represents the dissociation curve you would expect under conditions in the body.

3.4 Meeting the body's oxygen requirements

When the rate of respiration increases, haemoglobin releases more oxygen to the respiring cells. There are two mechanisms responsible for this:

1 The greater the rate of respiration, the more oxygen a cell consumes and the lower the partial pressure of oxygen. Look again at the curve in Figure 3.6. As the partial pressure of oxygen falls, so does the percentage saturation of the haemoglobin. In other words, reducing the amount of oxygen present in the cell increases the amount of oxygen that the blood will give up.

2 The greater the rate of respiration, the greater the amount of carbon dioxide that is produced. This pushes the dissociation curve even further to the right, again encouraging the haemoglobin to release more of the oxygen that it is carrying.

Don't forget that reducing the partial pressure of oxygen will increase the amount of **oxygen** that haemoglobin gives up to the cells.

3.5 Some different dissociation curves

Different organisms have slightly different sorts of haemoglobin. One way in which they differ is in the sequence of amino acids that makes up the globin part of the molecule. This sequence of amino acids affects the oxygen-carrying properties of the molecule. Figure 3.8 shows an oxygen dissociation curve for human haemoglobin and for three other types of haemoglobin.

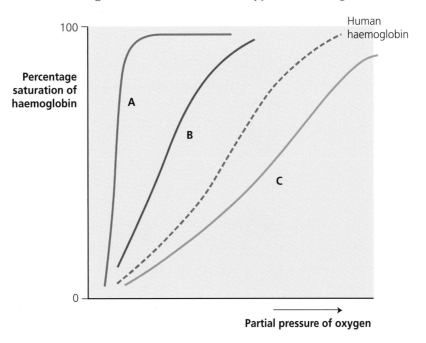

Figure 3.8 Different types of haemoglobin

Curve A

Haemoglobin with a dissociation curve that has this shape usually acts as an oxygen store. The haemoglobin holds on to its oxygen and only releases it when the amount of oxygen in the tissues falls to a very low concentration.

Examples of animals with this sort of haemoglobin include diving mammals such as seals and whales. Their tissues contain **myoglobin**, a haemoglobin-like pigment. Myoglobin releases its oxygen when the partial pressure of oxygen in the muscles falls to a very low level during lengthy dives.

Curve B

This type of haemoglobin can be saturated with oxygen even when there is only a limited amount available.

Examples of animals with this type of haemoglobin include the llama which lives high in the mountains of South America and the human fetus inside the uterus of its mother. They both live in environments with relatively low partial pressures of oxygen.

Curve C

This type of haemoglobin gives up its oxygen readily. It is associated with animals that have a high rate of respiration such as birds and small mammals such as shrews.

3.6 Transporting carbon dioxide

Carbon dioxide is transported in the blood mainly in the form of hydrogen carbonate ions (Figure 3.9).

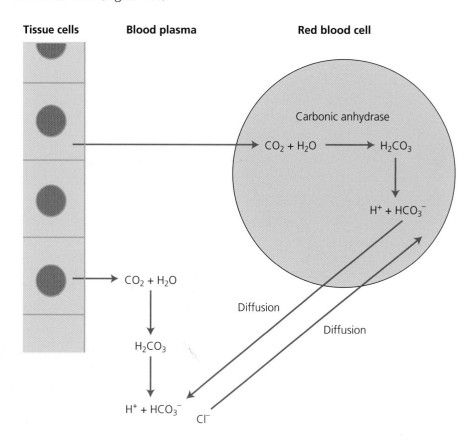

Figure 3.9 The transport of carbon dioxide

- Some carbon dioxide diffuses from respiring tissues into the blood plasma where it is converted to carbonic acid. Carbonic acid dissociates to produce hydrogen (H^+) ions and hydrogencarbonate (HCO_3^-) ions. In the plasma, this is a fairly slow reaction.
- Carbon dioxide also diffuses into the red blood cells and forms carbonic acid. This is a much faster reaction because it is catalysed by an enzyme, **carbonic anhydrase**, which is found in the red blood cells.
- There is a higher concentration of hydrogencarbonate ions in the red blood cells than in the plasma. Hydrogencarbonate ions diffuse from the cells into the plasma.
- Chloride (Cl^-) ions diffuse from the plasma into the red blood cells. This maintains a neutral charge.
- The hydrogen ions are taken up by buffers in the plasma and by haemoglobin which acts as a buffer in the red blood cells.
- When the blood reaches the lungs, the reverse series of reactions takes place.

A red blood cell placed in a concentrated sodium chloride solution will shrink as it loses water by osmosis. This has nothing to do with the chloride shift.

This is a good test. Try drawing a similar diagram to Figure 3.9 showing what happens to the carbon dioxide when blood reaches the lungs.

4 *The circulation of blood*

When you have finished revising this topic, you should be able to:

- distinguish between a single and a double circulation
- relate the structure of blood vessels to their functions
- explain the role of capillaries in the formation of tissue fluid
- explain how substances required for metabolism are supplied to cells and how waste products are removed

4.1 Single and double circulations

A **single circulation** is one in which blood passes through the heart once in its passage round the body. Fish have **single circulations** (Figure 3.10).

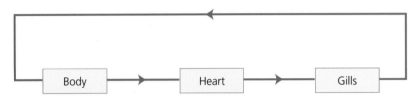

Figure 3.10 The single circulation of a fish

- Blood leaves the heart and passes to the gills where it is oxygenated. Oxygenated blood then goes to the rest of the body and finally returns to the heart.
- The heart has a single atrium and a single ventricle.
- Blood reaching the organs of the body have already been through the capillary network in the gills. It is therefore at a relatively low pressure.

In a **double circulation**, blood flows through the heart twice in its passage round the body. Mammals have a double circulation (Figure 3.11).

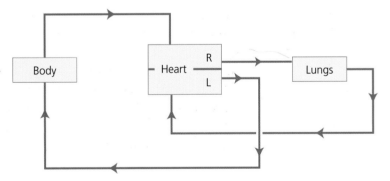

Figure 3.11 The double circulation of a mammal
- Blood goes from the right side of the heart to the lungs, where it is oxygenated, and back to the left side of the heart. This is the **pulmonary circulation**. In the **systemic circulation**, the oxygenated blood is pumped from the left side of the heart to the body.

- The heart has four chambers, a right atrium and ventricle and a left atrium and ventricle. The left ventricle has thicker muscle than the right ventricle. This results in higher blood pressure in the systemic circulation than in the pulmonary circulation. The volume of blood pumped out by the left side of the heart, however, is the same as that pumped out by the right side.
- Blood reaching the organs of the body are at a much higher pressure than that reaching the body organs of an animal with a single circulation. High pressure results in the rapid flow of blood. This helps to maintain a steep diffusion gradient which results in more efficient exchange of substances between the blood and the body tissues.

Both single and double circulations involve blood transporting substances in bulk from one exchange surface to another. Because of this, the blood system is sometimes referred to as a **mass transport system**.

4.2 Blood vessels

Blood leaves the heart in **arteries**. It flows through **arterioles** into the **capillaries** where exchange of substances with the tissues takes place. Blood from the capillaries is collected by the **venules**. It flows back to the heart in the **veins**.

Arteries and veins have a similar structure. Their walls consist of three layers:
1 An outer layer, the **tunica adventitia**, which contains collagen fibres.
2 A middle layer, the **tunica media**, which contains elastic fibres and smooth muscle. When the ventricles contract and force blood into the arteries, they bulge outwards, stretching the elastic fibres. Recoil of these elastic fibres helps to smooth out the flow of blood and force it along the arteries. Contraction of smooth muscle in the walls of the smaller arteries and arterioles ensures that the amount of blood flowing to the organs of the body can be varied according to need.
3 An inner layer, the **tunica intima**, which is made up of thin epithelial cells. These are smooth and reduce friction between the blood and the vessel walls.

Table 3.3 shows some of the differences between arteries, veins and capillaries.

> A vein is a blood vessel which returns blood to the heart. Use the word only in this context. Use the term blood vessel if you want to refer to either an artery or a vein.

> Blood will only flow from a high pressure to a low pressure. The blood pressure must therefore be higher in capillaries than in veins.

Feature	Artery	Capillary	Vein
Outer layer of wall	Present	Absent	Present
Middle layer of wall	Very thick with many elastic fibres	Absent	Present but thinner than in artery wall
Inner layer of wall	Present	Present	Present
Blood pressure	Highest (10–16 kPa)	Medium (2–4 kPa)	Lowest (less than 1 kPa)
Size of lumen	Relatively small	Approximately the same diameter as a red blood cell	Relatively large
Valves	Absent	Absent	Present

Table 3.3

4 The circulation of blood

4.3 Capillaries and tissue fluid

A capillary wall is only one cell thick. It is not the same as a cell wall. Use these terms correctly.

Table 3.3 shows that capillaries have only one wall layer. This consists only of thin epithelial cells. There are very small gaps between these cells which allow substances to pass out through the capillary wall. Look at Figure 3.12. It shows part of a capillary.

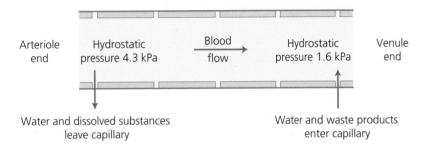

Figure 3.12 The formation of tissue fluid

- The **hydrostatic pressure** of the blood entering the capillary has a value of approximately 4.3 kPa. It results from contraction of the ventricles of the heart.
- Blood at the arteriole end of the capillary contains dissolved substances including proteins (called **plasma proteins**). As a result, water tends to move into the capillary by osmosis.
- There is a greater tendency for the hydrostatic pressure to force water out than for osmosis to draw it back in. As a result, water and dissolved substances leave the capillary at the arteriole end. This forms the **tissue fluid**.
- Because fluid is lost from the capillary, the hydrostatic pressure falls until it reaches approximately 1.6 kPa at the venule end.
- The plasma protein molecules are too large to leave the capillary, so water will still tend to move back into the blood by osmosis.
- There is now a greater tendency for water to move in by osmosis than to be forced out by the hydrostatic pressure of the blood. Water and waste products therefore re-enter the capillary.
- More fluid leaves the capillaries than is reabsorbed into them. This excess fluid drains into the lymph capillaries which join into larger **lymph vessels**. The lymph vessels eventually empty into the veins in the neck.

4.4 Supplying the tissues

The cells of the body are bathed in **tissue fluid**. Tissue fluid is rather like a bath with the taps running and the plug out. The bath may be full of water, but the water is always changing. Tissue fluid is also continually being formed and reabsorbed. Freshly formed tissue fluid has a high concentration of substances such as oxygen, glucose and mineral ions needed by cells. These substances pass into cells in the usual way, mainly by diffusion and active transport. Similarly, waste products pass out of the cells and into the tissue fluid.

5 *The mammalian heart*

When you have finished revising this topic, you should be able to:

- identify the main features of a mammalian heart
- describe how a heart beat is coordinated
- describe the heart cycle and relate the main events to changes in volume and pressure
- explain how valves prevent backflow of blood

5.1 Introduction

The heart functions as a pump maintaining the flow of blood. In order to understand how it works, you will need to be able to identify its main features. These are labelled on the simplified diagram shown in Figure 3.13.

> It may seem very simple, but many candidates confuse the right and left sides of the heart. Sort them out!

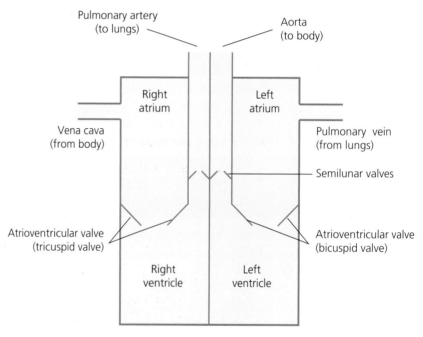

Figure 3.13 A simplified diagram of the heart

5.2 Coordinating the heart beat

Think about a tube of toothpaste. Suppose you wanted to get all the toothpaste out through the nozzle as fast as possible. There would be little point in holding it round the middle and squeezing hard. The best approach would be to squeeze up from the bottom and the toothpaste would come out from the nozzle in the way you wanted. When the heart beats, blood must also flow through it in a controlled way, in through the atria and out through the ventricles.

Heart muscle does not need to be stimulated by a nerve before it will contract. The heart beat originates in the muscle itself and, for this reason, it is described as being **myogenic**. It starts in a small area of tissue in the wall of the right atrium called the **sinoatrial node (SAN)** and spreads through specialised muscle cells as shown in Figure 3.14. Again, we have used the simplest drawing possible to illustrate the structure of the heart.

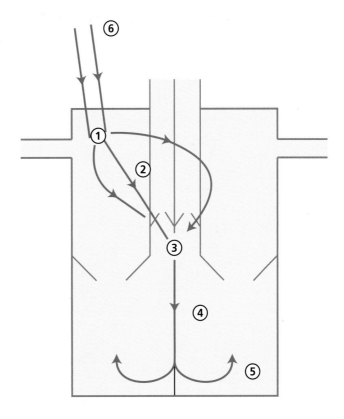

Figure 3.14 Coordinating the heart beat

1 The heart beat starts with an electrical signal at the sinoatrial node, sometimes called the pacemaker. The sinoatrial node has its own rhythm, but this may be modified by nerve impulses from the brain.

2 The electrical signal spreads through muscle cells in the walls of the atria. It causes the muscle to contract and blood to be forced from the atria into the ventricles.

3 The signal can only pass from the atria to the ventricles through one area of tissue, the **atrioventricular node (AVN)**. The signal slows down here. This allows time for the blood in the atria to be squeezed through into the ventricles before they contract.

4 The signal now passes rapidly down the specialised fibres which form the **bundle of His** in the wall or septum separating the two ventricles.

5 Finally, it spreads up through the muscle of the ventricle wall. This means that the ventricle contracts from the bottom upwards, forcing blood upwards and out through the pulmonary artery and the aorta.

6 The rate at which the sinoatrial node sends signals can be altered by nerve impulses coming from the brain. Impulses in the **sympathetic nerve** speed up the heart rate. Impulses from the **parasympathetic** or **vagus nerve** slow the heart rate.

5.3 The heart cycle

The events described above lead to a cycle of events involving the filling and emptying of the heart. Although it is a continuous cycle, it can be conveniently divided into the three main parts shown in Figure 3.15.

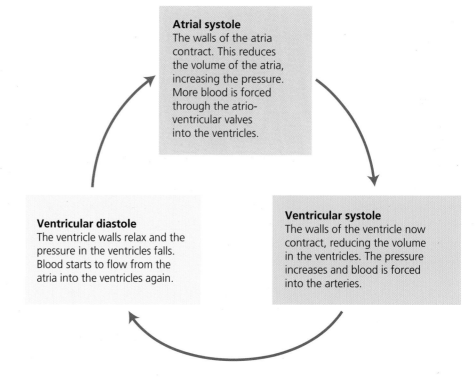

Figure 3.15 The heart cycle

5.4 The heart valves and their role

Imagine you are standing in a shower with an unopened umbrella. Hold the umbrella up. The pressure of the water from above will keep the umbrella closed and water will flow past. Now turn the umbrella upside down. The pressure will force it open and prevent the water from flowing past.

We can apply these principles to the valves in the heart. Look again at Figure 3.13 and remind yourself of how the valves are arranged:

- When the pressure is higher in the atria than in the ventricles, the atrioventricular valves will open and allow blood to flow from the atria to the ventricles.
- When the pressure is higher in the ventricles than in the atria, the atrioventricular valves close preventing blood flowing back into the atria.
- When the pressure is higher in the ventricles than in the arteries leaving the heart, the semilunar valves open, allowing blood to flow into the arteries.
- When the pressure is higher in the arteries than in the ventricles, the semilunar valves close preventing blood flowing back into the ventricles.

6 *Transport in plants*

When you have finished revising this topic, you should:

- ■ understand that xylem and phloem form mass transport systems in plants
- ■ be able to explain how the cohesion-tension theory can account for the flow of water through the xylem
- ■ be able to describe how transpiration takes place and explain how variation in environmental factors may affect the rate of transpiration

Atrioventricular might seem like an unnecessarily complex term, but it tells you precisely where the atrioventricular valves are situated. Call them left and right atrioventricular valves and you won't have to worry about which is the bicuspid, and which is the tricuspid.

Think about valves opening and closing in terms of pressures. If you do this you can usually work out when a valve will be shut and when it will be open.

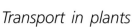
- be able to explain how water and mineral ions can enter and move through a root
- be able to explain how mass transport can account for the movement of sugars through phloem

6.1 Introduction

The transport systems of plants are similar to those of animals in that they also rely on a combination of diffusion and active transport at exchange surfaces and mass transport to distribute these substances around the organism.

Plants have two main transport systems:

1 **Xylem** takes water and mineral ions from the roots to the stems and leaves.
2 **Phloem** takes sugars and other organic substances from where they are formed in the leaves to where they are needed in the developing shoots, flowers and fruits, and in the roots.

Xylem transports substances up the stem to the aerial parts, but phloem can transport substances both upwards and downwards.

Although these two systems have different functions, some transfer of substances does occur between the xylem and the phloem, particularly in the younger parts of the plant.

6.2 Movement in xylem

The cells that transport water and mineral ions through a plant are dead. They have lost their end walls and fit together rather like a series of drain-pipes to form **vessels**. The walls of vessels are thickened with **lignin**. This makes them impermeable. There are three mechanisms which contribute to the movement of water through these xylem vessels.

Capillarity

Xylem vessels form a series of very fine tubes from the roots up to the leaves. They range in diameter between 20 and 400 μm. When a tube of this diameter is dipped in water, the liquid will move upwards through capillary action. Capillary action helps to explain how water rises up the stem, but the pressure produced is not enough to account for a rise of more than about a metre.

Root pressure

Some plants are able to transfer mineral ions into the xylem in their roots by active transport. This produces a water potential gradient. Water enters the xylem from the surrounding tissues by osmosis and is forced up the stem by root pressure. Root pressure is not enough to account for the movement of water to leaves at the top of trees.

Cohesion-tension

This is the most important of the three mechanisms. The argument can be broken into four main steps:

1 Leaves **transpire**. Water evaporates from the moist conditions inside a leaf, through the stomatal pores to the drier surrounding air.
2 Water molecules demonstrate a property known as **cohesion**. Because of hydrogen bonds that form between neighbouring water molecules, they stick

It might seem very obvious but make sure you are quite clear about the basic differences between phloem and xylem: where they are located in a stem and what they do.

Transport in plants depends to a large extent on differences in water potential and osmosis. Make sure you have revised these important concepts first.

to each other. This means that as water is lost by transpiration, more is pulled up the xylem to replace it.

3 The pulling action of transpiration stretches the water column in the xylem so that it is under tension.

4 Water molecules also cling to the walls of the xylem. This is **adhesion** which also helps to pull the water column upwards.

It is useful to compare what happens in the xylem with what happens when you stretch an elastic band. Pulling both results in tension. Use the elastic band model to explain evidence for cohesion-tension.

6.3 The factors affecting transpiration

Table 3.4 summarises information about the way in which some factors can affect the rate of transpiration.

Factor	Explanation
Factors affecting the supply of water Dry conditions Increased salinity of soil	Both of these factors lead to a small water potential gradient between the soil and the root cells. The movement of water into the root will be reduced. In some cases, water may even move out of the plant into the soil.
Damage to the roots	If the roots are damaged by pests, less water will enter the plant.
Factors affecting the stomata Light intensity	Light intensity has an effect on the opening and closing of stomata. Since approximately 90% of the water lost in transpiration passes through the stomatal pores, light intensity will have an important effect on the rate of transpiration.
Factors affecting water loss from the plant Temperature	An increase in temperature increases the random movement of molecules, leading to a faster rate of diffusion.
Wind speed Relative humidity	Both of these factors affect the water potential gradient between the inside of the leaf and the outside. The smaller this gradient, the slower the rate of water loss.

Table 3.4

6.4 Passage through the roots

Use the terms 'mineral ions' or 'inorganic ions' rather than 'nutrients'. Nutrients is too vague.

Most mineral ions are in a lower concentration in the soil surrounding the roots than they are in the cells. They are therefore taken into the roots by active transport.

The higher concentration of mineral ions inside the root cells means that a water potential gradient exists. Water moves by osmosis from the higher (less negative) water potential in the soil to the lower (more negative) water potential in the cells.

Once inside the root, water and mineral ions can move from cell to cell by two main pathways:

1 The **apoplastic pathway** is the way in which most water moves. This pathway involves movement through the cell walls and the spaces between cells.

2 The **symplastic pathway**. The contents of neighbouring cells are joined by thin strands of cytoplasm which pass through the cell walls. These strands are

called **plasmo-desmata**. Substances can move through the plasmo-desmata and pass from one cell to another without having to cross cell membranes. This is probably the most important pathway for the transport of mineral ions.

6.5 Movement in phloem

Phloem is made up from several types of cell. Those through which sugars and other substances are transported are arranged in columns called **sieve tubes**. They differ from xylem vessels in a number of ways. These are summarised in Table 3.5.

Xylem	Phloem
No cross walls between cells	Cross walls between cells form **sieve plates** which are perforated with a number of small pores
Cell walls thickened with lignin	Cell walls not thickened with lignin
Dead cells with no cytoplasm	Contain cytoplasm but no nucleus and very few organelles

Table 3.5

There are several theories which attempt to explain transport in phloem, none of which is completely convincing. Figure 3.16 summarises one of these theories, the **mass flow hypothesis**. Carbohydrates, mainly in the form of sucrose, are loaded into the phloem at the **source**, perhaps a photosynthesising leaf. They are transferred to a **sink**, a root or other part of the plant that requires sugars.

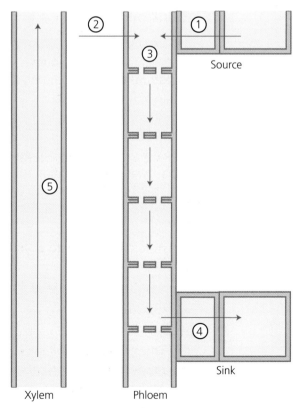

Figure 3.16 Transport in phloem, the mass flow hypothesis

1 Sugars are produced in actively photosynthesising leaf cells. They are loaded into the sieve tube by active transport.

2 The resulting increase in solute concentration lowers the water potential in the sieve tube. Water moves by osmosis from the higher water potential outside to the lower water potential inside.

3 The inflow of water sets up a hydrostatic pressure in the sieve tube, forcing the water and the solutes that it contains through the sieve tube to the sink.

4 At the sink, the sugars are removed from the phloem and transferred to the surrounding cells where they are either used in respiration or stored as starch. The water potential outside the sieve tube will now be lower than that inside, and water will leave the phloem by osmosis.

5 Xylem is responsible for transporting the water back up the stem.

7 *The digestive system*

When you have finished revising this topic, you should be able to:

■ identify the different stages in the processing of food in the gut of a mammal

■ describe the role of digestive juices in the digestion of carbohydrates, lipids and proteins

■ explain how the products of digestion are absorbed

■ explain how the structure of the small intestine is adapted to its functions of digestion and absorption

■ explain how the secretion of digestive juices is controlled

7.1 Introduction

There are four main stages involved in the processing of food in the gut of a mammal:

1 **Ingestion.** Food is taken into the mouth and mechanically broken down and crushed by the teeth.

2 **Digestion.** The process in which large insoluble molecules are broken down into smaller soluble products by the action of various digestive enzymes. Digestive enzymes are **hydrolases**. They break down chemical bonds by adding water. Carbohydrates are broken down to glucose and other monosaccharides, triglycerides to glycerol and fatty acids, and proteins to amino acids.

3 **Absorption.** The products of digestion are absorbed into the body through the gut wall.

4 **Egestion.** The elimination of undigested food, dead cells and bacteria as faeces.

7.2 Digestion

Carbohydrates

Carbohydrate digestion is summarised in Figure 3.17.

You should distinguish between digestion and absorption. If a question asks for a description of digestion, you do not need to discuss absorption as well.

Egestion is the removal of material that has never been absorbed into the body. Excretion is the removal of waste that has been produced in metabolic reactions in the body.

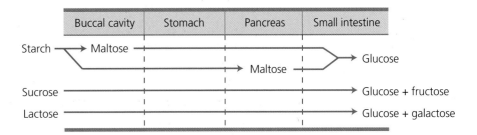

Figure 3.17 The digestion of carbohydrates

1 Saliva contains **amylase** which hydrolyses some starch to the disaccharide, maltose. The amount of amylase in saliva is often related to the amount of starch in the diet.

2 Pancreatic juice also contains amylase. This completes the digestion of starch to maltose. A second amylase is necessary because salivary amylase is inactivated by the acid conditions in the stomach.

3 The final stage of starch digestion takes place in the small intestine. The **maltase** enzymes that hydrolyse maltose to glucose are found in the plasma membranes of the epithelial cells lining the small intestine.

4 The enzymes **sucrase** and **lactase** which hydrolyse sucrose and lactose are also found in the plasma membranes of the epithelial cells.

Triglycerides

Triglyceride digestion is summarised in Figure 3.18.

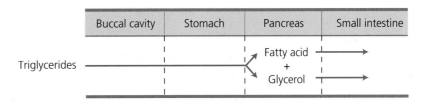

Figure 3.18 The digestion of triglycerides

When mixed with water, triglycerides, such as the fats in the diet, form large globules. A large globule of fat has a relatively small surface area to volume ratio and will, therefore, only be broken down slowly by enzymes. Bile is added to food in the upper part of the small intestine. Although it does not contain enzymes, it helps in the digestion of triglycerides in two ways:

1 Bile salts help to keep fats **emulsified**. The action of the muscles in the wall of the intestine breaks globules of fat down into much smaller droplets. Bile salts prevent these small droplets from joining together again into a bigger drop. This increases the surface area for enzyme action.

2 Bile is alkaline. It helps to create the optimum pH for lipase to digest triglycerides.

Lipase is secreted by the pancreas. It hydrolyses triglycerides forming glycerol and fatty acids.

Although bile is important in digestion, it does not contain any digestive enzymes.

The pancreas is a gland which secretes digestive juice into the small intestine. Food does not pass through it.

Proteins

Before looking at protein digestion in detail, we need to distinguish between the two main types of protein-digesting enzyme:

1 **Endopeptidases** hydrolyse peptide bonds between specific amino acids in the middle of the polypeptide chains that make up protein molecules. They break large polypeptide chains into smaller ones.

2 **Exopeptidases** hydrolyse peptide bonds between amino acids at the end of polypeptide chains. They produce a mixture of amino acids, dipeptides and tripeptides.

Protein digestion is summarised in Figure 3.19.

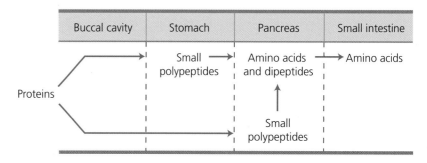

Figure 3.19 The digestion of proteins

1 **Pepsin** is an endopeptidase secreted by the stomach. It works most effectively in acid conditions such as those found in the stomach. Pepsin is secreted in an inactive form called **pepsinogen**. The hydrochloric acid produced by the stomach together with pepsin already formed convert pepsinogen to pepsin.

$$\text{pepsinogen} \xrightarrow[\text{hydrochloric acid}]{\text{pepsin}} \text{pepsin}$$

2 The pancreas contains a number of different protein-digesting enzymes. These include:

- Another endopeptidase called **trypsin**. Trypsin works best in alkaline conditions. It is also secreted in an inactive form, **trypsinogen**, which is converted to trypsin by the enzyme, **enterokinase**, produced in the wall of the small intestine.

$$\text{trypsinogen} \xrightarrow{\text{enterokinase}} \text{trypsin}$$

- Exopeptidase enzymes which break off groups of one, two or three amino acids from the ends of the polypeptides.

3 The final stage of protein digestion takes place in the small intestine. Dipeptides and tripeptides are digested by enzymes found in the plasma membranes and the cytoplasm of the epithelial cells lining the small intestine.

Table 3.6 summarises some of the more important enzymes involved in digestion.

Unfortunately, the information in this table has to be learnt. If you know it, you can get some very easy marks; if you don't, you can get into a dreadful mess.

Part of gut	Secretion	Enzymes	Substrate	Product(s)
Buccal cavity	Saliva	Amylase	Starch	Maltose
Stomach	Gastric juice	Pepsin	Polypeptides	Smaller polypeptides
Pancreas	Pancreatic juice	Amylase	Starch	Maltose
		Lipase	Triglyceride	Glycerol + fatty acids
		Trypsin	Polypeptides	Smaller polypeptides
		Exopeptidases	Polypeptides	Dipeptides + tripeptides
Wall of small intestine		Maltase	Maltose	Glucose
		Sucrase	Sucrose	Glucose + fructose
		Lactase	Lactose	Glucose + galactose
		Dipeptidase	Dipeptides	Amino acids

Table 3.6

7.3 Absorption

Some small molecules such as water and ethanol can be absorbed through the stomach wall, but most of the absorption of the products of digestion takes place in the small intestine:

- **Amino acids and proteins.** Amino acids are water soluble and are absorbed partly by diffusion and partly by active transport. Although most proteins are digested in the gut, some are absorbed intact. New-born babies, for example, are able to absorb protein antibodies from milk.
- **Glucose.** Glucose is also absorbed by a mixture of diffusion and active transport. The absorption of both glucose and amino acids is linked to the active transport of sodium ions from the gut.
- **Water.** Active transport of sodium ions, amino acids and glucose across the wall of the intestine creates a water potential gradient. Water moves by osmosis from a higher water potential in the gut to a lower water potential outside.
- **Fatty acids and glycerol.** Fatty acids and glycerol molecules combine with bile salts to form small drops called **micelles**. The micelles dissolve in the cell surface membrane and diffuse into the cytoplasm of the epithelial cells. Triglycerides are reformed and packaged with protein and cholesterol to form **chylomicrons**. The chylomicrons pass along lymph vessels into the blood. The blood of a person who has recently eaten a fatty meal contains large numbers of chylomicrons.

7.4 Structure and function

As food is moved along the small intestine, absorption takes place. Movement is the result of waves of muscle contraction called **peristalsis**. The wall of the small intestine is surrounded by two muscle layers, an outer layer of longitudinal muscle and an inner layer of circular muscle:

- The longitudinal muscle contracts and the circular muscle relaxes in front of the ball of food that is being pushed along the intestine. This makes the cavity of the intestine wider, allowing food to be pushed into it.
- Immediately behind the ball of food, the circular muscle contracts and the longitudinal muscle relaxes. This squeezes the food forwards.

The wall of the small intestine forms a very efficient absorption surface because it is specially adapted:

It has a large surface area

- The lining is folded and these folds are covered in finger-like projections called **villi**.
- The epithelial cells that form the lining of the villi have plasma membranes that are further folded into **microvilli**. This extensive folding gives a very large surface area over which diffusion and active transport can take place.

It maintains a large difference in concentration on either side of the exchange surface

- Muscles in the gut wall bring about movements which ensure that the contents of the small intestine are constantly moving. This brings more products of digestion into contact with the epithelial cells.
- The circulation of blood through the capillaries in the villi ensures that blood containing high concentrations of glucose and amino acids is replaced with blood low in concentration of these substances. This helps to maintain the diffusion gradient across the exchange surface.
- Active transport is involved as well as diffusion. The epithelial cells contain large numbers of mitochondria. These supply the necessary ATP.

It has a thin exchange surface

- Although the epithelial cells are tall and columnar in shape, there is only a short distance between the lumen of the gut and the blood in the capillaries.

7.5 Controlling digestive secretions

Digestive juices are only secreted in large amounts when food is present in the relevant part of the gut. Apart from wasting the body's resources, prolonged contact of these juices with the gut wall could damage the tissues involved. Table 3.7 summarises the way in which digestive secretions are controlled.

In looking at the adaptations of the small intestine that ensure efficient absorption, you can use the framework suggested for gas exchange (page 51).

Make sure that you distinguish between villi and microvilli.

Part of gut	Stimulus	Effect
Buccal cavity	Contact of substances in food with taste buds	Reflex leading to secretion of saliva.
	Various stimuli associated with food	Conditioned reflex leading to secretion of saliva.
Stomach	Contact of substances in food with taste buds and various stimuli associated with food	Reflex and conditioned reflex leading to secretion of gastric juice.
	Presence of food in the stomach	The hormone **gastrin** is released from cells in the stomach wall. It stimulates the secretion of gastric juice.
Pancreas and gall bladder	Presence of food in the small intestine	The hormone **pancreozymin-cholecystokinin (CCKPZ)** is released from cells in the intestinal wall. It stimulates the release of bile by the gall bladder and enzymes by the pancreas.
		The hormone **secretin** is released from the cells of the intestinal wall. It stimulates the release of alkaline fluid by the pancreas.

Table 3.7

1 Defining and measuring biodiversity

When you have finished revising this topic, you should:

- be able to define what is meant by the term biodiversity
- describe ecosystems characterised by low and high biodiversity
- understand how biodiversity can be estimated in an ecological context

1.1 Defining biodiversity

Biodiversity refers to the variety and complexity of life and may be considered at different levels. It can be measured for example within a habitat or at a genetic level.

Biodiversity is often taken as an indicator of the number of different species present in an ecosystem. Ecosystems with a large diversity of species tend to be more stable (resistant to change) than those that are less diverse.

- Extreme ecosystems like tundra, deserts and salt marshes are characterised by low diversity. In areas of low diversity, plant and animal populations are mainly affected by **abiotic** factors.
- Ecosystems with high diversities are usually mature, natural (i.e. not created by human activity) and have environmental conditions that are not too hostile. In these ecosystems, populations are mostly affected by **biotic** factors.

1.2 Measuring biodiversity

Ecologists tend to look at three key factors when examining the diversity of an ecosystem.

- Species frequency — how abundant a species is in an area.
- Species richness — the total number of different species in an area.
- Percentage cover (plants only) — how much of the surface is covered by a particular plant species.

The **diversity index (d)** is a measure of the diversity of an ecosystem. It is calculated using the following formula:

$$d = \frac{N\,(N-1)}{\Sigma n\,(n-1)}$$

where N = total number of organisms of all species, n = total number of one species, and Σ = the sum of $n(n-1)$. Ecosystems with greater diversity have a higher diversity index.

2 Classification

When you have finished revising this topic, you should:

- understand the process of taxonomy
- be able to list the key taxonomic groups in order of increasing specificity
- describe the key features of the five kingdoms used in classification

2.1 Taxonomy

Taxonomy is the process by which living organisms are organised into groups. The most useful classification system is based on the evolutionary relationships between organisms and has a hierarchical structure. All organisms belong to one of five **kingdoms**: Animalia, Fungi, Plantae, Prokaryotae and Protoctista.

Each kingdom can be subdivided into a number of phyla. Each phylum can also be divided into classes, and so on. The smallest group is an individual species. Table 4.1 shows how humans are classified.

Kingdom	Animalia
Phylum	Chordata
Class	Mammalia
Order	Primates
Family	Hominidae
Genus	Homo
Species	Sapiens

Table 4.1 Classification of humans

Each organism is known by its genus and species, according to the binomial naming system used in taxonomy (the process of classification). So humans are known as *Homo sapiens* and polar bears are known as *Ursus maritimus*.

Cladistics is a method of taxonomy that focuses on the features of organism that are evolutionary developments. This emphasis on phylogeny (the genetic relation between organisms) is useful as it enables us to determine points in time when one species split into two, such as when humans and chimpanzees split from a common ancestor about 6 million years ago.

2.2 Animals and fungi

Animalia is the kingdom containing animals. All animals share the following features:
- they are multicellular organisms
- their cells do not possess cell walls
- they show heterotrophic nutrition

Examples include starfish (members of the phylum Echinodermata) and lizards (members of the phylum Chordata).

Fungi is the kingdom containing mushrooms, toadstools, moulds and yeasts. All fungi share the following features:
- they are heterotrophic
- they reproduce by means of spores
- they are usually made up of thread-like structures known as hyphae
- most have cell walls made of chitin

Examples include *Mucor* and yeast.

Animals and fungi are eukaryotic, multicellular and heterotrophic. The key difference between these kingdoms at a cellular level is that fungus cells have cell walls (made of chitin) and animal cells do not.

2.3 Plants, protoctists and prokaryotes

Plantae is the kingdom containing plants. All plants share the following features:
- they are multicellular organisms
- their cells possess cell walls composed mainly of cellulose
- they show autotrophic nutrition
- they have a life cycle which shows alternation of generations

Examples include mosses (members of the phylum Bryophyta) and ferns (members of the phylum Filicinophyta) and flowering plants (members of the phylum Angiospermophyta).

Protoctista is the kingdom containing eukaryotic organisms that cannot be classified as members of the Fungi, Animalia or Plantae kingdoms. Some have a cell wall and some do not. Examples: red algae, slime moulds and amoeba.

Prokaryotae is the kingdom that includes bacteria. These organisms have a cell wall. Prokaryotic cells have a different structure to all other (eukaryotic) cells. The main differences are summarised in Table 2.2 on page 36.

Plants and protoctista are eukaryotic. Protoctista is an unusual kingdom in that it contains all the species that do not fit into the other eukaryotic kingdoms. Prokaryotes are unicellular and differ from all the other kingdoms due to their cellular structure.

3 | *Environmental adaptations*

When you have finished revising this topic, you should:
- be able to explain the nature of structural, physiological and behavioural adaptations and the differences between them
- be able to describe different types of environmental adaptations
- understand that organisms become adapted to their environments through the process of natural selection

3.1 Structural adaptations

Structural adaptations refer to any adaptations of the size or shape of an organism, and any specialised internal or external structures. For example, fish have fins and a streamlined body to help them swim efficiently in water. They also have gills, which are specialised for gas exchange.

3.2 Physiological adaptations

Physiological adaptations refer to any adaptations in terms of how the cells and metabolic processes function to benefit an organism. For example, plants growing in areas of high salinity have specially adapted cells containing very high salt concentrations to allow water uptake by osmosis.

3.3 Behavioural adaptations

Behavioural adaptations refer to any adaptations in terms of how an organism acts to make the most of its surroundings. For example, some animals hibernate in winter to avoid the low temperatures and lack of food.

3.4 Examples of adaptations

Xerophytes are plants that live in arid (dry) environments and show adaptations to conserve water. For example, they may have thicker, waxy cuticles on leaves and stems, leaves that have been reduced to spines (to reduce the surface area for water loss), or leaves that roll up in particularly dry conditions (to trap moist air, thereby slowing down transpiration).

Desert mammals show adaptations to living in a hot, dry climate. For example, kangaroo rats are nocturnal (being active during the cooler nights), produce concentrated urine (to reduce water loss) and are efficient at using the metabolic water produced by aerobic respiration.

Invertebrates have physiological and structural adaptations to cope with their environment's specific oxygen concentration level. For example, invertebrates demonstrate structural adaptations to maximise the surface area to volume ratio of respiratory surfaces. Others may possess respiratory pigments, such as haemoglobin for the efficient internal transport of oxygen.

Organisms vary in terms of how well they are adapted to their environment. Those that are better adapted are more likely to survive and reproduce, passing on the genes that enabled them to flourish. This process of **natural selection** will result in the gradual development of a population better adapted to the environment — unless that environment changes.

A2
Biology

CHAPTER 5 Energy for biological processes

1 *Energy transfer*

When you have finished revising this topic, you should:

- be able to explain the difference between catabolic and anabolic reactions
- be able to describe the part played by ATP in cell metabolism
- understand the relationship between photosynthesis and respiration in a leaf

1.1 Catabolism and anabolism

Metabolism is a term used to describe all the chemical reactions taking place in an organism. These reactions are divided into:

- **Catabolic reactions** in which larger molecules are broken down to smaller ones with the release of energy. Respiration involves a series of catabolic reactions.
- **Anabolic reactions** in which smaller molecules are built up into larger ones. This is energy requiring. Photosynthesis and protein synthesis are anabolic reactions.

There is a balance between catabolic and anabolic reactions. Catabolism provides the energy for the organism to synthesise larger molecules in its anabolic reactions.

It is best to write about the transfer of energy, although it is sometimes more convenient for biologists to refer to energy being released. Never write about reactions **making** energy.

1.2 ATP

Adenosine triphosphate, **ATP**, is a nucleotide. It consists of three components:

- a 5-carbon sugar — ribose
- an organic base — adenine
- three phosphate groups

Figure 5.1 shows how these components are arranged to form an ATP molecule.

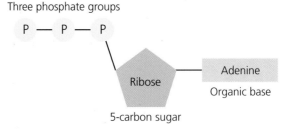

Figure 5.1 A molecule of ATP

ATP breaks down to form ADP and a phosphate group (written as P_i for inorganic phosphate). This breakdown involves hydrolysis of the ATP molecule, and releases energy that is used for energy-requiring reactions. Since more energy is always released than is required for the reaction, some is lost as heat.

To maintain the organism's anabolic reactions, ATP must be continually formed from ADP and phosphate. This reaction needs a source of energy. The formation and breakdown of ATP are summarised in Figure 5.2.

Energy transferred from catabolic reactions
such as respiration

$$ADP + P_i \rightleftharpoons ATP$$

Energy transferred to anabolic reactions
+
Energy lost as heat

Figure 5.2 The formation and breakdown of ATP

ATP is more useful than glucose as an immediate source of energy because:

- The breakdown of ATP makes energy instantly available. It takes longer for energy to be made available from the series of reactions involved in the breakdown of glucose during respiration.
- The breakdown of a molecule of ATP releases a small amount of energy ideal for driving an anabolic reaction. The breakdown of a molecule of glucose would produce much more energy than would be required.

1.3 Photosynthesis and respiration

Plants produce sugars from carbon dioxide and water. This process is anabolic and requires light energy. The sugars are used as the starting point for other anabolic reactions, such as those producing proteins, lipids and nucleic acids. The energy required for these reactions comes from ATP synthesised during respiration.

Although very basic, the relationship between photosynthesis and respiration in a leaf is frequently misunderstood:

- In daylight, the cells of the leaf are photosynthesising and respiring. The rate of photosynthesis is greater than the rate of respiration. If it were not, the plant would not be able to accumulate organic molecules and grow. Therefore, in the light there is a net uptake of carbon dioxide.
- At night, no light is available so the leaf cannot photosynthesise. It still respires. In the dark, there is a loss of carbon dioxide.

This relationship is summarised in Figure 5.3.

During the day the concentration of carbon dioxide in the intercellular spaces of a leaf decreases because it is used in photosynthesis. It does not increase because it is needed for photosynthesis.

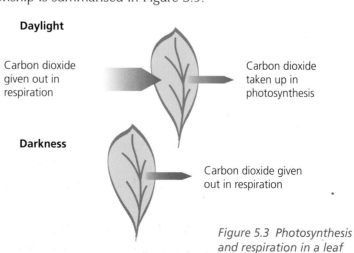

Daylight

Carbon dioxide given out in respiration

Carbon dioxide taken up in photosynthesis

Darkness

Carbon dioxide given out in respiration

Figure 5.3 Photosynthesis and respiration in a leaf

2 *Respiration*

When you have finished revising this topic, you should:

- understand that respiration involves the release of energy from organic molecules
- be able to explain how a molecule of glucose is oxidised to pyruvate with a net gain of ATP and reduced NAD
- know that pyruvate combines with coenzyme A to produce acetyl-coenzyme A
- be able to explain the role of the Krebs cycle in producing ATP and reduced coenzyme
- be able to explain how ATP is produced via a chain of electron carriers in the process of oxidative phosphorylation
- be able to describe the biochemical pathways of anaerobic respiration
- be able to calculate and interpret respiratory quotients

2.1 Introduction

> Distinguish between respiration and gas exchange. Respiration is a biochemical pathway. Gas exchange involves exchange of respiratory gases with the environment.

Respiration is often represented by the equation:

$$C_6H_{12}O_6 + 6O_2 \rightarrow 6CO_2 + 6H_2O + \text{energy}$$

This equation can be misleading. $C_6H_{12}O_6$ represents glucose, but glucose is not the only substance which can be used as a respiratory substrate. Similarly, under some circumstances, oxygen is not necessary for respiration, and water and carbon dioxide are only produced in the proportions shown in the equation when glucose is respired in the presence of oxygen. The only constant feature is the release of energy, and it is best to think of respiration as the biochemical pathway which takes place in cells and results in the release of energy from organic molecules.

At A2, we look in detail at only one of the many respiratory pathways. This is summarised by the equation above and represented in Figure 5.4.

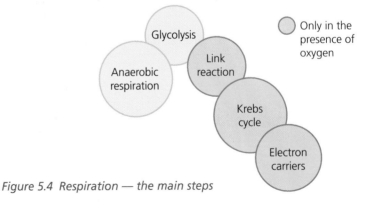

Figure 5.4 Respiration — the main steps

If you are not a chemist, the detail in the biochemical pathway of respiration may seem daunting. What you need to do is make sure that you have a sound understanding of the overall process. Once you have the general picture you can look in more detail at the individual steps in the process.

Glycolysis

The part of the biochemical pathway of respiration in which a molecule of glucose is broken down into two 3-carbon pyruvate groups. The boxes represent the number of carbon atoms present in the molecules and ions shown.

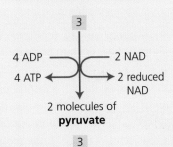

1 molecule of
glucose

| 6 |

2 ATP
2 ADP

Phosphate groups are added to the glucose molecule. These phosphate groups are supplied by ATP.

2 molecules of
glyceraldehyde 3-phosphate

| 3 |

4 ADP — 2 NAD
4 ATP — 2 reduced NAD

2 molecules of
pyruvate

| 3 |

The resulting phosphorylated sugar is then broken down to give 2 molecules of pyruvate. The process produces 4 ATP molecules so there is a net gain of 2 ATP for each molecule of glucose. The complete reaction is an oxidation reaction and releases hydrogen. This is removed and used to reduce a coenzyme known as NAD.

The link reaction

A name given to describe the reaction linking glycolysis to the Krebs cycle.

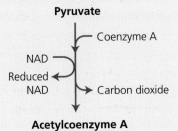

Pyruvate

Coenzyme A
NAD
Reduced NAD — Carbon dioxide

Acetylcoenzyme A

In the link reaction, pyruvate combines with coenzyme A to produce acetyl-coenzyme A. This reaction involves the loss of one molecule of carbon dioxide. It is also an oxidation reaction and the hydrogen that is lost is used to reduce NAD.

Krebs cycle

A series of oxidation reactions involving the release of carbon dioxide which leads to the production of ATP and reduced coenzyme.

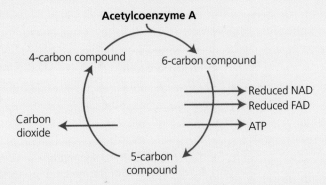

Acetylcoenzyme A

4-carbon compound   6-carbon compound

Reduced NAD
Reduced FAD

Carbon dioxide   ATP

5-carbon compound

Acetylcoenzyme A can be thought of as a 2-carbon compound. It is fed into Krebs cycle where it combines with a 4-carbon compound to produce a 6-carbon compound. The 6-carbon compound is broken down in a series of oxidation reactions to produce the 4-carbon compound again. Krebs cycle involves the loss of carbon dioxide and the production of ATP and reduced coenzymes.

Electron carrier systems
A chain of molecules in which the energy released in the passage of electrons from one molecule to the next is used to produce ATP.

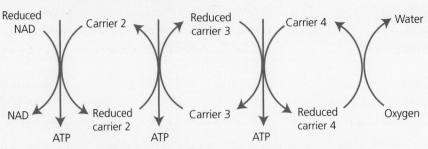

Hydrogen released during Krebs cycle acts as a source of electrons and protons. The electrons are passed from molecule to molecule along the electron transport chain. At each transfer a small amount of energy is released. This is used to pump protons through the inner mitochondrial membrane on which the carriers are situated. When the protons return through the membrane they release energy which is used to produce ATP.

2.2 Anaerobic respiration

In some circumstances, there is too little oxygen available to enable a cell to respire by the pathway described above. In order to produce ATP under these conditions, it must respire anaerobically. Anaerobic respiration relies on glycolysis to produce ATP, so it is far less efficient than aerobic respiration. A single molecule of glucose can produce 38 molecules of ATP in aerobic respiration; it only produces 2 molecules of ATP in anaerobic respiration.

During glycolysis, NAD is reduced. This reduced NAD is normally reconverted to NAD by the reactions of the aerobic pathway. If the cell is unable to respire aerobically, there would come a time when all the NAD would be reduced and glycolysis would be unable to continue. In anaerobic respiration, pyruvate is converted either to lactate in animals, or to ethanol and carbon dioxide in plants and in microorganisms such as yeast. This allows NAD to be reformed from reduced NAD. These pathways are summarised in Figure 5.5.

> Remember the end product of anaerobic respiration in animals is lactate, not ethanol. Marathon runners don't fall down drunk!

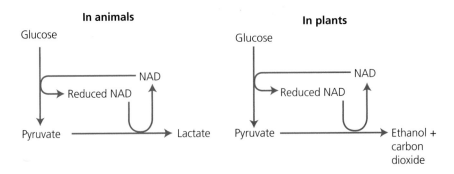

Figure 5.5 The biochemical pathways of anaerobic respiration

2.3 Using different respiratory substrates

The respiratory substrate is the organic substance which is required. Oxygen is not a respiratory substrate.

The organic substance which forms the starting-point for respiration is called a **respiratory substrate**. One way in which it is possible to find out which substrate is being respired is by calculating the **respiratory quotient** or **RQ**. The respiratory quotient is the amount of carbon dioxide produced in a given time divided by the amount of oxygen consumed in the same time.

RQ is easy to calculate providing that you remember which way up the equation is.

Calculating the respiratory quotient

The equation below represents the respiration of a triglyceride (a fat):

$$2C_{51}H_{98}O_6 + 145O_2 \rightarrow 102CO_2 + 98H_2O$$

$$RQ = \frac{\text{amount of carbon dioxide produced}}{\text{amount of oxygen consumed}}$$

Using the equation representing the respiration of the triglyceride,

$$RQ = \frac{102}{145}$$

$$= 0.7$$

Table 5.1 shows RQs for some important respiratory substrates.

RQ	Respiratory substrate
0.7	Triglyceride
0.9	Protein
1.0	Carbohydrate

Table 5.1

We have to be careful about interpreting RQ. An RQ of 0.9, for example, may mean that the respiratory substrate is protein, but a mixture of triglyceride and carbohydrate could also give the same value.

3 *Photosynthesis*

When you have finished revising this topic, you should be able to explain that:

- photosynthesis involves light-dependent reactions in which light energy is captured by chlorophyll and used to produce ATP and reduced NADP

- the light-dependent reactions also involve photolysis, in which water molecules break down to produce electrons, hydrogen ions and oxygen

- carbon dioxide is reduced to carbohydrates in the light-independent reactions

3.1 Introduction

Photosynthesis is similar to respiration in that it is also a complex process involving a number of separate stages. It is useful to look at the overall process shown in Figure 5.6 before considering the detailed biochemistry.

The coenzyme involved in photosynthesis is NADP; that involved in respiration is NAD (P for photosynthesis).

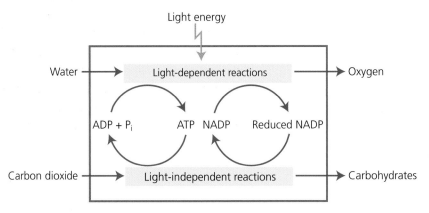

Figure 5.6 *Photosynthesis, the main steps*

There are two basic steps:

1 The **light-dependent** reactions in which light energy is captured by chlorophyll and is used to produce ATP and reduced NADP.
2 The **light-independent** reactions in which the ATP and reduced NADP are used in the conversion of carbon dioxide to carbohydrate.

It is convenient to consider these reactions separately.

3.2 The light-dependent reactions

Figure 5.7 summarises the light-dependent reactions.

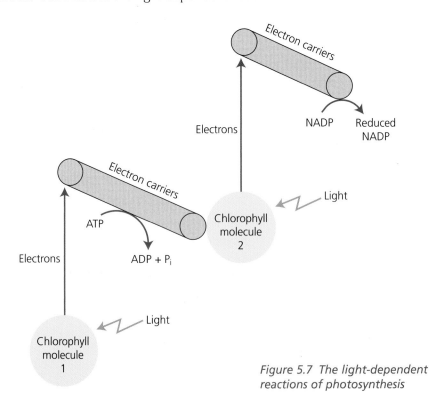

Figure 5.7 *The light-dependent reactions of photosynthesis*

1 Light strikes the first chlorophyll molecule (called chlorophyll molecule 1 for convenience). It excites some of the electrons in the molecule. These are raised to a higher energy level and pass to a molecule which acts as an electron acceptor.

2 These electrons are transferred along the series of molecules which form an electron carrier system. Energy is released and used to produce ATP.

3 The electrons are eventually accepted by chlorophyll molecule 2.

4 Light strikes chlorophyll molecule 2 and some of its excited electrons also pass down an electron carrier system.

5 Eventually these electrons are used to convert NADP to reduced NADP.

6 Another reaction is also involved. This is photolysis. Water is broken down to produce protons, electrons and oxygen. This is summarised by the equation:

$$2H_2O \quad \rightarrow \quad 4H^+ \quad + \quad 4e^- \quad + \quad O_2$$

| water | protons | electrons | oxygen |

- The protons help to reduce the NADP to reduced NADP.
- The electrons are accepted by chlorophyll molecule 1.
- The oxygen is given off as a waste product.

3.3 The light-independent reactions

This is the process in which carbon dioxide is reduced to form carbohydrates. It is summarised in Figure 5.8.

Light-independent reactions are so called because they are independent of light. However, they cannot take place without the ATP and reduced NADP produced in the light. Because of this, the light-independent reactions cannot continue for long in the dark.

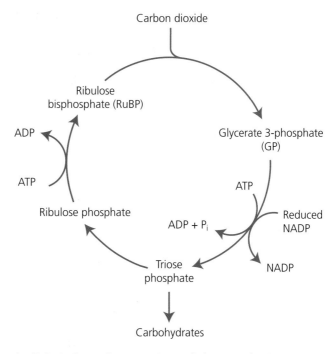

Figure 5.8 The light-independent reactions of photosynthesis

Remember, the
Calvin cycle is part
of photosynthesis;
the Krebs cycle is part
of respiration.

1 Ribulose bisphosphate (RuBP) is a 5-carbon compound. It combines with a molecule of carbon dioxide to form two molecules of the 3-carbon compound, glycerate 3-phosphate (GP).

2 GP is reduced to triose phosphate (also known as glyceraldehyde 3-phosphate, GALP). This requires reduced NADP and energy from ATP.

3 Some of the triose phosphate is converted into carbohydrates such as glucose and starch.

4 The rest of the triose phosphate is used to make more RuBP in a cycle of reactions known as the Calvin cycle. ATP is also required for the regeneration of RuBP. It supplies the phosphate necessary to convert ribulose phosphate to ribulose bisphosphate.

CHAPTER 6

Cellular control

1 Nucleic acids

When you have finished revising this topic, you should:

- be able to draw a simple diagram of a nucleotide and explain how nucleotides join to form nucleic acids
- be able to describe the structure of DNA
- know how the structure of DNA differs from that of mRNA and tRNA

1.1 Nucleotides

Nucleic acids are polymers. The basic unit, or monomer, from which they are all formed, is a **nucleotide**. Figure 6.1 shows the structure of a nucleotide.

This is a simplified diagram, but it is all you need to know.

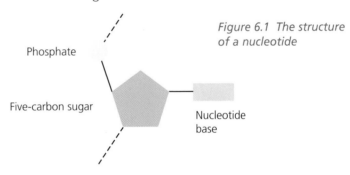

Phosphate

Five-carbon sugar

Nucleotide base

Figure 6.1 The structure of a nucleotide

A nucleotide has three components which are linked by condensation reactions. The components are:

1 **A phosphate group.**
2 **A 5-carbon or pentose sugar.** In DNA, this sugar is deoxyribose; in RNA, it is ribose.
3 **A nitrogen-containing base.** There are five different bases: adenine, cytosine, guanine, thymine and uracil. These are usually abbreviated to A, C, G, T and U. The nucleotides that make up a molecule of DNA contain one of the four bases A, C, G or T. The nucleotides that make up a molecule of RNA contain one of the four bases A, C, G or U. There are two different sorts of base: purines and pyrimidines. Adenine and guanine are purines; cytosine, thymine and uracil are pyrimidines. Table 6.1 summarises this information.

Unless you have to name the nucleotide base concerned, it is quite acceptable to abbreviate it to the first letter. If you have to name a base, make sure that you spell it correctly. There are a number of other biological words which are very similar to adenine, for example.

Purines and pyrimidines. You ought to know which is which.
Try: **Pure silver.**
The chemical symbol for silver is Ag, so **A**denine and **g**uanine are **pur**ines.

A phrase that might help you to remember a few more facts about nucleotide bases is:
A tea for two. Adenine pairs with thymine — **A** and **T**. There are **two** hydrogen bonds between them.

Base	Type of base		Found in	
	Purine (contains two rings of atoms)	**Pyrimidine (contains one ring of atoms)**	**DNA**	**RNA**
Adenine	✔		✔	✔
Cytosine		✔	✔	✔
Guanine	✔		✔	✔
Thymine		✔	✔	
Uracil		✔		✔

Table 6.1 Summary of nucleotide bases

1.2 The structure of DNA

If we link a large number of nucleotides together with condensation reactions, we get a **polynucleotide strand**. A molecule of DNA consists of two polynucleotide strands joined together by hydrogen bonds. Figure 6.2 shows part of a DNA molecule.

A simple but very important point: DNA is not a protein. Proteins are built up from amino acids; nucleic acids are built up from nucleotides.

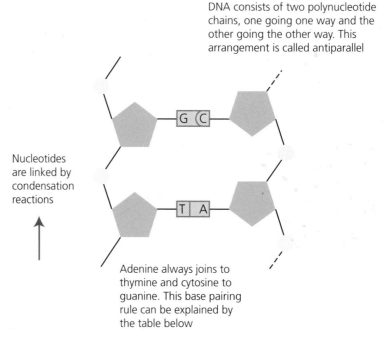

DNA consists of two polynucleotide chains, one going one way and the other going the other way. This arrangement is called antiparallel

Nucleotides are linked by condensation reactions

Adenine always joins to thymine and cytosine to guanine. This base pairing rule can be explained by the table below

Base	Type	Number of hydrogen bonds	Type	Base
Adenine	Purine	Two	Pyrimidine	Thymine
Guanine	Purine	Three	Pyrimidine	Cytosine

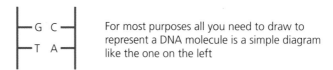

For most purposes all you need to draw to represent a DNA molecule is a simple diagram like the one on the left

Figure 6.2 The structure of DNA

1.3 RNA and its structure

The structure of an RNA molecule differs from that of a molecule of DNA in a number of ways. These include the following:

- An RNA molecule consists of one, not two, polynucleotide strands. This single strand may be twisted.
- RNA contains the 5-carbon sugar, ribose. It does not contain deoxyribose, the sugar found in DNA.
- RNA does not contain the nucleotide base, thymine. Instead it contains uracil.

There are three types of RNA which are found in cells.

Messenger RNA (mRNA)

Messenger or mRNA acts, as its name suggests, as a messenger molecule. It takes a copy of the genetic code from the DNA in the nucleus to the ribosomes in the cytoplasm during protein synthesis.

Transfer RNA (tRNA)

tRNA, like all the other types of RNA, contains the base uracil, not thymine.

During protein synthesis, the amino acids which make the protein must be arranged in the correct order. The function of transfer or tRNA is to collect the amino acids from the cytoplasm and assemble them in the correct order on the mRNA molecule. Figure 6.3 shows the main features of a tRNA molecule.

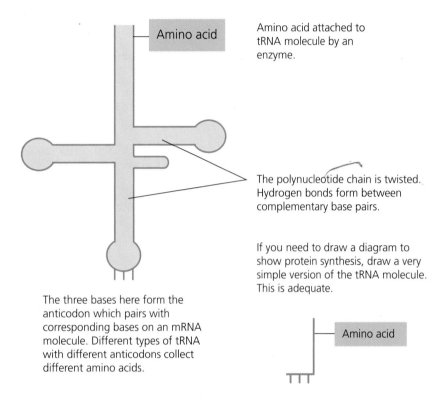

Figure 6.3 A transfer RNA molecule

Ribosomal RNA (rRNA)

Ribosomes are very small organelles made of protein and RNA. This RNA is known as ribosomal or rRNA.

1.4 Differences between nucleic acids

A-level questions often require you to describe the differences between the different types of nucleic acid molecules. Some of these are shown in Table 6.2 (overleaf).

	DNA	mRNA	tRNA
Nucleotides contain the base thymine	✔		
Nucleotides contain ribose		✔	✔
Complementary bases joined with hydrogen bonds	✔		✔
Single polynucleotide chain		✔	✔

Table 6.2 Some of the differences between types of nucleic acid molecules

2   *The genetic code and protein synthesis*

When you have finished revising this topic, you should be able to:

- describe the main features of the genetic code and explain how a gene codes for a specific polypeptide or protein
- describe the main types of gene mutation and explain how mutations can lead to changes in the proteins produced by cells
- describe how mRNA is produced from DNA in the process of transcription
- describe how translation of mRNA results in the formation of a specific polypeptide or protein

2.1 The genetic code

> Nucleic acids are made up from nucleotides; proteins are made up from amino acids.

A gene is a piece of DNA which has a specific function, usually to code for a particular polypeptide or protein. In section 1.2, we saw that a DNA molecule consists of two polynucleotide strands. The strand which actually carries the code is called the **coding strand**. The other strand, the **non-coding strand**, carries the complementary base sequence. It is the sequence of bases on the coding strand of the DNA molecule that forms the genetic code. It determines the sequence of amino acids that will make up the protein for which it codes. The genetic code has a number of features.

It is a triplet code

> Make sure that you can calculate the total number of combinations of 2 and 3 bases.

There are four different bases in a DNA molecule: adenine (A), cytosine (C), guanine (G) and thymine (T). There are 20 different amino acids which form proteins. Clearly, one base cannot code for one amino acid. What about two bases coding for each amino acid? There are rather more different combinations of two bases: AA, AC, AG, AT, CA, CC, and so on. If you work out all the possible combinations of two bases, you get 4×4, 4^2 or 16 — still not enough to code for 20 different amino acids. Suppose then we try three bases. This will give us 4^3 or 64 different combinations, more than enough to code for all 20 amino acids.

It is a degenerate code

There are clearly more base combinations than there are amino acids. This means that several base sequences may code for the same amino acid. Let's take an example (there is no need to learn it). The DNA base sequences CCA, CCC, CCG and CCT, all code for the same amino acid: glycine. The genetic code

is therefore described as being **degenerate**. There are two important consequences of this:

1 Nearly all amino acids have more than one code.

2 The first two bases of the code are more important than the third base in specifying a particular amino acid.

The code is non-overlapping

Each sequence of three bases codes for a different amino acid. Again, we will look at a specific example. The DNA sequence CCACAT, codes for two amino acids: glycine (CCA) and valine (CAT). An individual base does not appear in more than one coding sequence. In other words, the codes do not overlap.

It is a universal code

The same base sequence always codes for the same amino acid. The sequence CCA, for example, codes for glycine in humans, in buttercups, and in bacterial cells. It is because the code is universal that genetic engineering is possible and a gene transferred from one species to a completely different one will still produce the same protein.

2.2 Gene mutations

A **gene mutation** is a change in one or more of the nucleotide bases in the DNA of an organism. This changes the organism's genotype, which means that mutations may be inherited. Mutations occur randomly, and may affect any point in an organism's DNA. The rate at which mutation occurs may be increased by exposure to **mutagens** such as ultra-violet radiation, X-rays, and various organic substances.

Table 6.3 shows a number of different types of mutation. Some of these have little if any effect on the organism concerned; others have major effects.

When explaining features concerning the genetic code, it is useful to be able to quote an actual example. You can make up a base sequence; you don't need to know which amino acid is coded for by which base sequence.

Type of mutation	Effect on genetic code	Base sequence
No mutation	DNA not affected	CGA GGC GTT AAG
Mutations that have little effect on the organism		
Substitution	One base replaced by another	CG**T** GGC GTT AAG
Mutations that have a major effect on the organism		
Duplication	Repetition of part of the base sequence	CG**C G**AG GCG TTA AG
Deletion	Removal of part of the base sequence	GAG GCG TTA AG
Addition	Addition of extra base	CG**T** AGG CGT TAA G

Table 6.3 Types of mutation

Mutations that have little effect on the organism

In a **substitution**, one base is substituted for another. In the example shown in Table 6.3, the third base in the sequence, adenine has been replaced by thymine. At the most, this will only affect one amino acid, the one normally coded for by the base sequence CGA. The rest of the genetic code has not been altered in any way.

Moreover, because the genetic code is a degenerate code, it is possible that altering the third base will have no effect whatsoever on the protein produced. This is because the mutated sequence will code for exactly the same amino acid as the original sequence. This is true in this example. CGA and CGT — both code for the amino acid, alanine.

Mutations that have a major effect on the organism

Deletion involves removing a base. Look at the example in Table 6.3. Here, the first base, cytosine, has been lost. This results in a change to all the following codes. Look at each group of three bases; they have all been altered. The first three bases in the sequence are now GAG, not CGA, so they will code for a different amino acid. The removal of one base has therefore affected all the following codes. This effect of shifting the whole sequence is sometimes referred to as a **frame shift**. Duplication and addition also produce frame shifts.

2.3 Protein synthesis

Protein synthesis consists of two stages:

1 The base sequence on the coding strand of the DNA molecules is copied or **transcribed** to produce a molecule of mRNA.

2 This mRNA moves out of the nucleus to a ribosome where its base sequence is read or **translated** to form a polypeptide.

This is summarised in Figure 6.4.

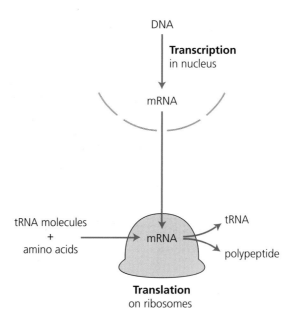

Figure 6.4 Transcription and translation

Transcription

The main steps in transcription are summarised in the following flow chart.

Don't confuse transcription and translation. It might help to remember that **c** comes before **l** in the alphabet.

Make sure that you distinguish between transcription in which a molecule of mRNA is produced, and DNA replication in which more molecules of DNA are made.

The hydrogen bonds between the complementary bases in the two DNA strands that make up a particular gene break. The strands separate. This process is controlled by enzymes and other proteins.

↓

The coding strand of the DNA now forms a template for the formation of an mRNA molecule. Individual RNA nucleotides found in the nucleus line up against the complementary bases on the coding strand of the DNA.

↓

The RNA nucleotides now join together to make an mRNA molecule. This moves out through the pores in the nuclear envelope to a ribosome.

Translation

The second stage of protein synthesis is translation, in which the base sequence of the mRNA is read or translated and a polypeptide is formed. This is summarised in Figure 6.5.

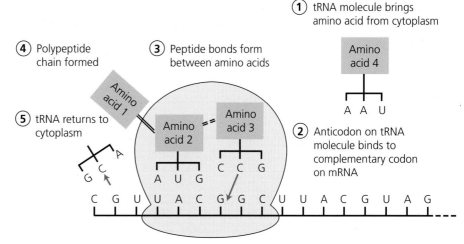

Figure 6.5 Translation

Genetic manipulation

When you have finished revising this topic, you should be able to describe the biological principles underlying each of the following:

- the use of genetic engineering to produce proteins from microorganisms
- the use of genetic fingerprinting to show differences between DNA obtained from different individuals
- the use of the polymerase chain reaction to make large amounts of DNA from very small samples

3.1 Introduction

Genetic manipulation is an aspect of biotechnology which involves investigating or altering the DNA of an organism. It is an area of biology in which progress

There are many examples of genetic engineering producing useful products. Concentrate on learning about one or two of them. It is much better to learn about what has been done rather than speculating on what might be achieved in the future.

has been extremely rapid. For this reason, this section concentrates on general principles rather than specific examples.

3.2 Proteins from microorganisms

Genes may be extracted and introduced into microorganisms which can produce the products in large amounts. Many substances are now made in this way but the essential steps in the process are similar. These are summarised in the following flow chart.

Learn what each of the enzymes used in genetic engineering does and remember, they work in exactly the same way as other enzymes.

The gene is isolated. Possible ways of isolating a gene are:

- Part of the base sequence that forms the gene can be located using a **DNA probe**. This is a single strand of DNA which, under the right conditions, binds to the complementary base sequence and identifies the required gene. **Restriction enzymes** are used to cut the DNA on either side of the gene, producing a length of DNA containing this gene.

- The gene can be produced from the relevant mRNA molecule. An enzyme known as **reverse transcriptase** is used to make a single strand of complementary DNA or cDNA. From this, a double-stranded copy of the gene can be produced.

- It is possible to work backwards from the amino acid sequence that you want to produce. A piece of DNA with the appropriate base can be synthesised.

↓

Transferring the gene to the microorganism

Once you have isolated the gene for the protein you wish to produce, the next stage is to introduce it into a suitable host microorganism, such as a bacterium. For this, a **vector** is used. A vector is really a carrier. One of the most commonly used vectors is a **plasmid**. This is a small circular piece of DNA which is found in bacterial cells. Plasmids are isolated and then cut open using the same restriction enzyme that was used to cut the gene from the original DNA. This results in the same sequence of bases on the cut ends and allows the gene to join up with the DNA in the plasmid. Another enzyme, **DNA ligase**, is used to join these pieces of DNA together.

The plasmids containing the new gene are known as **recombinant plasmids**. The next step is to get the recombinant plasmids into the bacterial cells. This is not very successful, and usually only about 1 in 40 000 bacterial cells will contain the recombinant plasmid. We therefore need a way to check that we have the right bacterial cells before moving on to the next stage in the process. One thing we can do is to introduce two genes into the plasmid, the one we want to use and one which makes the recombinant plasmid easy to detect. Genes that code for antibiotic resistance are often used for this purpose.

↓

Making the product

The bacterial cells containing the required gene are now grown in a fermenter. Under ideal conditions they multiply rapidly and reproduce to form genetically identical **clones**, all of them containing the required gene and producing the required product.

3.3 Genetic fingerprinting

Genetic fingerprinting is the name given to a technique which is used to compare samples of DNA. A lot of the DNA found in the nucleus of a cell does not actually code for proteins. In fact, we are not really sure what it does. Some of this non-coding DNA consists of short sequences of bases that may be repeated many times. The number of repeats varies from person to person. Genetic fingerprinting involves a comparison of the similarities and differences in these pieces of DNA. The flow chart below shows how this technique can be used to identify evidence left at the scene of a crime.

DNA is isolated from a suitable sample. This may be from blood or semen. It could even be from cells sticking to a cigarette end.

↓

The DNA is cut into smaller pieces using restriction enzymes. Some of these pieces will contain the repeated sequences that are being investigated. Their actual length will depend on the number of times the particular base sequence has been repeated. If it is relatively few times, the piece of DNA concerned will be small. If, on the other hand, the base sequence has been repeated may times, the piece of DNA will be larger.

↓

Electrophoresis is used to separate the pieces. The DNA is placed on a layer of gel and an electric current applied. The DNA pieces separate out according to their size and electrical charge, with the smaller pieces travelling further than the larger ones. The end result is that there are bands of DNA separated out like rungs on a ladder.

↓

Bands of DNA containing the repeated sequences are now identified using a **DNA probe**. This is a single strand of DNA which attaches to and identifies the repeated base sequence. The position of the bands can be used to compare different individuals.

3.4 The polymerase chain reaction

This is an extremely useful technique that enables us to produce large amounts of DNA from a very small sample by copying it many times. The main steps in the process are shown in the following flow chart.

The DNA sample is heated to a temperature of 95°C.
This separates it into its two strands.

The sample is mixed with DNA nucleotides, the enzyme
DNA polymerase, and a primer. The primer is a short piece of DNA
which acts as a signal telling the enzyme where to start copying.
The temperature is reduced to 40°C. At this temperature, the
primer will bind to the piece of DNA which is being copied.

The temperature is raised to 70°C. DNA polymerase is thermostable
and will now copy each strand of the original DNA.
At the end of this part of the reaction there will be
two molecules of DNA, each identical to the original one.

This cycle can then be repeated, each time doubling
the amount of DNA.

4   *From one generation to the next*

When you have finished revising this topic, you should be able to:

- describe the behaviour of chromosomes during meiosis
- explain the importance of meiosis in maintaining the chromosome number and in contributing to genetic variation
- list the main differences between mitosis and meiosis

4.1 Meiosis

Meiosis is the name given to a type of nuclear division in which the number of chromosomes is halved. Because of this, it is usually associated with sexual reproduction. We will look at the same simple organism that we used to demonstrate mitosis in section 3.2 of chapter 2. You may remember that there were two pairs of chromosomes in each of its body cells. One of the chromosomes from each pair originally came from the male parent, while the other came from the female parent.

Meiosis involves two separate stages, usually referred to as meiosis I and meiosis II:

1 In **meiosis I**, the chromosomes come together in their homologous pairs with each chromosome split longways into two **chromatids**. One member of each pair of chromosomes is now pulled to each end of the cell so that, at the end of this stage, there will be one complete set of chromosomes at one pole of the cell and one complete set at the other. Some of the chromosomes in each set will have come originally from the male parent and some will have come originally from the female parent.

The differences might seem obvious but there is a lot of confusion between mitosis and meiosis. Try not to revise these topics one after the other.

Don't worry about crossing over at this stage. Concentrate on the movement of the chromosomes.

2 In **meiosis II**, each of the daughter cells formed from the first stage will divide again. Each chromosome consists of two chromatids, and these chromatids will be pulled apart as the cells divide. This is shown in Figure 6.6.

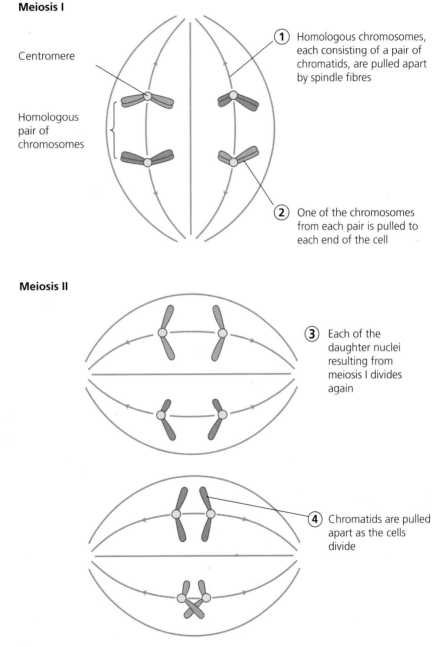

Meiosis I

Centromere

Homologous pair of chromosomes

(1) Homologous chromosomes, each consisting of a pair of chromatids, are pulled apart by spindle fibres

(2) One of the chromosomes from each pair is pulled to each end of the cell

Meiosis II

(3) Each of the daughter nuclei resulting from meiosis I divides again

(4) Chromatids are pulled apart as the cells divide

Figure 6.6 Meiosis

During the first division of meiosis, another event may take place which also gives rise to genetic variation. This is **crossing over**. During the first stage of meiosis, the pairs of homologous chromosomes have come together. The four chromatids involved twist round each other. During this process, an individual chromatid may break and rejoin to a different one. If one of the chromatids involved originally came from the male parent and the other from the female parent, there could be a new mixture of genetic material on the chromatids concerned. Crossing over is extremely important, as it produces genetic variation. The process of crossing over is shown in Figure 6.7.

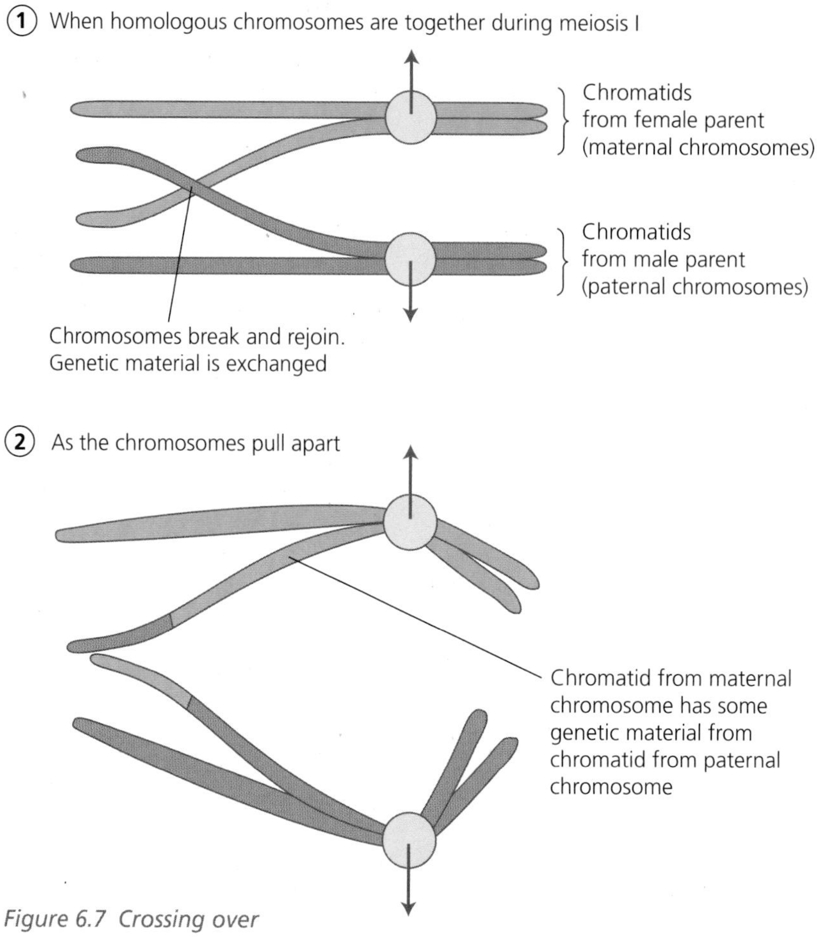

① When homologous chromosomes are together during meiosis I

}Chromatids from female parent (maternal chromosomes)

}Chromatids from male parent (paternal chromosomes)

Chromosomes break and rejoin. Genetic material is exchanged

② As the chromosomes pull apart

Chromatid from maternal chromosome has some genetic material from chromatid from paternal chromosome

Figure 6.7 Crossing over

4.2 The importance of meiosis

Meiosis ensures that the chromosome number remains constant from one generation to the next. Think about the human life cycle. The body cells in an adult each contain 46 chromosomes. Meiosis takes place during the formation of the sex cells or **gametes**, resulting in each sperm cell and each egg cell containing 23 chromosomes, one chromosome from each pair. When fertilisation takes place, a sperm cell fuses with an egg cell and the full diploid number of chromosomes is restored.

Meiosis produces genetic variation. There are three features which lead to this:

1 Separation of daughter chromosomes during meiosis I. During the first stage of meiosis, these chromosomes are pulled apart and move to the two poles of the cell. Those originally coming from the female parent (the **maternal** chromosomes) and those originally coming from the male parent (the **paternal** chromosomes) behave independently of one another. Therefore, at the end of the first division, different daughter cells will have different mixtures of maternal and paternal chromosomes.

2 Separation of daughter chromatids during meiosis II. Because this is also a random event, different combinations of chromatids may appear in the nuclei created by the second division of meiosis.

3 Crossing over results in mixing of genetic material from maternal and paternal chromatids.

Other sexually reproducing organisms have similar life cycles, but you should note that meiosis does not always come at the same stage.

If you are asked to compare two things in a table, make sure you compare like with like.

Some haploid cells divide by mitosis, so it is not correct to say that mitosis always produces diploid cells while meiosis produces haploid ones.

4.3 Meiosis and mitosis

There are a number of basic differences between meiosis and mitosis. Some of these are summarised in Table 6.4.

Meiosis	Mitosis
The number of chromosomes is halved.	The number of chromosomes remains the same.
Meiosis involves two divisions and results in four daughter nuclei being formed.	Mitosis involves one division and results in two daughter nuclei being formed.
During the first division of meiosis, chromosomes come together in their homologous pairs, each pair appearing as a group of four chromatids.	During mitosis, the chromosomes do not come together in homologous pairs. Each chromosome consists of two chromatids.
Crossing over may result in genetic variation.	Crossing over may take place, but it does not result in genetic variation.

Table 6.4 Meiosis and mitosis

5 Genetics and genetics problems

When you have finished revising this topic, you should:

- understand the meanings of the basic terms used in genetics
- be able to solve genetics problems showing dihybrid inheritance
- be able to solve problems showing sex linkage

5.1 Basic terms

Here is a list of basic terms used in genetics. Make sure that you know what each of these terms means and get into the habit of using them correctly.

Gene
A length of DNA that occupies a particular place, known as a **locus**, on a chromosome. It has a specific sequence of nucleotide bases. This means that a gene has a specific function and often codes for a particular protein.

Allele
A particular form of a gene.

You should be careful to distinguish between genes and alleles. All humans have a gene coding for the protein that determines ABO blood groups. Different people, however, have different alleles of this gene so they can have different blood groups, A, B, AB or O.

Dominant
A dominant allele is one that is always expressed in the phenotype. In humans, the allele for freckles, **F**, is dominant to that for no freckles, **f**. Since the allele for freckles is dominant, individuals with either the genotype **FF** or **Ff** will have freckles.

Recessive

A recessive allele is only expressed in the phenotype when the other allele of the pair is identical. Since the allele for no freckles is recessive, only individuals with the genotype **ff** will have no freckles.

Codominant

Two different alleles are described as codominant if both are expressed in the phenotype.

Genotype

The genotype of an organism describes the alleles that it contains. The genotype of a person with freckles may be either **Ff** or **FF**.

Phenotype

The characteristics of an organism determined by its genes and its environment. Freckles require the presence of the dominant allele, **F**. They are made more prominent by exposure to sunlight, an environmental factor.

Heterozygote

An organism in which the alleles at a particular locus differ from each other. A human with the genotype **Ff** is a heterozygote.

Homozygote

An organism that has identical alleles at a particular locus. Human with the genotypes **FF** or **ff** are homozygotes.

5.2 Problems with dihybrid inheritance

The best way of revising for this topic is probably to work carefully through a straightforward example like the one following and then attempt other problems from past examination papers.

Example

*In guinea pigs, coat colour and hair length are determined by two genes situated on different pairs of chromosomes. The allele for black colour, **B**, is dominant to that for white colour, **b**, and the allele for short hair, **L**, is dominant to that for long hair, **l**. Show, by means of a genetic diagram, the expected results of a cross between two guinea pigs, heterozygous for these alleles.*

Parental phenotypes: Black, short hair   Black, short hair

Parental genotypes: **BbLl**   **BbLl**

Gametes: (BL)(Bl)(bL)(bl)   (BL)(Bl)(bL)(bl)

Offspring genotypes:

		Male gametes			
		(BL)	(Bl)	(bL)	(bl)
Female gametes	(BL)	BBLL	BBLl	BbLL	BbLl
	(Bl)	BBLl	BBll	BbLl	Bbll
	(bL)	BbLL	BbLl	bbLL	bbLl
	(bl)	BbLl	Bbll	bbLl	bbll

Offspring phenotypes:	Black, short hair	9
	Black, long hair	3
	White, short hair	3
	White, long hair	1

Some points to note:

- In this example you have been provided with the letters representing the alleles. Sometimes you have to choose your own. Try to choose a letter that relates clearly to one of the phenotypes. If possible, avoid the following: C, I, J, K, O, P, S, U, V, W and Z because the only difference between the capital and lower-case versions is one of size. In an examination, it is very easy to confuse them. The use of X and Y is best confined to the sex chromosomes.
- Set your answer out clearly. The method shown above is the one recommended by the Institute of Biology and used by all examination boards. Get into the habit of using it and the headings showing what each stage represents.
- Gametes are haploid. Each gamete contains one chromosome from each pair. In the example above, a gamete will therefore contain one of the alleles for colour and one of the alleles for hair length. Always draw a ring round your gametes and check that each one contains one of each pair of alleles. This is where most mistakes are made in simple problems of this sort.
- Set out your working, however simple, as a checkerboard or Punnett square. You are less likely to make mistakes than if you link alleles like a series of spider's webs.
- Check off each of the genotypes as you count it. It will be easier to find your mistake if you end up with a total of 15 or 17!

5.3 Problems with sex linkage

A gene is described as being sex-linked if it is found on one of the sex chromosomes. A sex-linked gene codes for the protein that enables humans to distinguish between red and green objects. The dominant allele of this gene, **R**, codes for the normal pigment, so a person possessing this gene has normal colour vision. The recessive allele, **r**, codes for the faulty pigment that produces colour blindness. We represent the dominant allele as X^R and the recessive allele as X^r. The letter X shows us that the allele is on the X chromosome.

Here is a straightforward problem involving sex linkage.

> **Example**
>
> *A man, who is red-green colour blind, marries a woman who is homozygous for the allele for normal colour vision. Show, by means of a genetic diagram, the possible genotypes and phenotypes of their children.*
>
> | Parental phenotypes: | Colour-blind man | Woman with normal vision |
> |---|---|---|
> | Parental genotypes: | **X^rY** | **X^RX^R** |
> | Gametes: | Ⓧʳ Ⓨ | Ⓧᴿ Ⓧᴿ |

Offspring genotypes:

		Male gametes	
		X^r	Y
Female gametes	X^R	$X^R X^r$	$X^R Y$
	X^R	$X^R X^r$	$X^R Y$

Offspring phenotypes: $X^R X^r$ Girl with normal vision

                            $X^R Y$ Boy with normal vision

Some points to note:

- In problems which show sex linkage, always show the X and Y chromosomes. Note that, in this example, the gene concerned is located on the X chromosome. The Y chromosome will not have one of these alleles. Represent the alleles in the way shown here.
- In this example, the girl is a heterozygote. She has normal colour vision but is able to pass the X^r allele on to her children. She is therefore sometimes referred to as a carrier. Her phenotype, however, is for normal vision.

6 Selection

When you have finished revising this topic, you should be able to:

- explain examples of selection in terms of a number of basic principles
- explain what is meant by directional, stabilising and disruptive selection
- understand the processes of allopatric and sympatric speciation

6.1 Introduction

Selection is a process where the best-adapted organisms in a population survive and reproduce, passing on their alleles to the next generation. There are many examples of selection but the underlying principles are the same in each case:

- The living organisms in a population vary.
- Some of this variation is genetic and originally arose as a result of mutation.
- Selection operates on this population. Some varieties have an advantage and some have a disadvantage.
- The varieties that have an advantage survive and reproduce.
- They pass on the favourable alleles to the next generation, ultimately leading to a change in the overall genetic make-up of the population.

Let us consider two examples and use these five principles to explain them. This is set out in Table 6.5.

Try this for yourself. Find another example of selection and make sure that you can explain it in terms of these five principles.

The evolution of bacteria which are resistant to a particular type of antibiotic	The development of cattle which produce large amounts of milk
1 In a population of bacteria, some will be resistant to a particular antibiotic while some will not be resistant.	In a population of cattle, some will produce larger amounts of milk than others.
2 The resistance to antibiotics is genetic and controlled by genes. Different alleles of these genes originally arose as a result of mutation.	Differences in milk yield are partly genetic and controlled by genes. Different alleles of these genes originally arose as a result of mutation.
3 When a patient is treated with the relevant antibiotic, only those bacteria which are resistant to it will survive.	Records will show which cows produce most milk. Only these cows will be kept for breeding.
4 The bacteria with the favourable alleles have an advantage. They survive and reproduce …	The cows with the favourable alleles have an advantage. They survive and reproduce …
5 … passing on these alleles to the next generation. Eventually this leads to a population of bacteria with a greater frequency of alleles for antibiotic resistance.	… passing on these alleles to the next generation. Eventually this leads to a population of cows with a greater frequency of alleles for high milk yield.

Table 6.5

6.2 The different results of selection

When we start with some aspect of a population that shows continuous variation between one extreme and the other, selection can operate in different ways.

Directional selection

This operates against one extreme. Suppose the *x*-axis of the graphs in Figure 6.8 represents hair length in house mice.

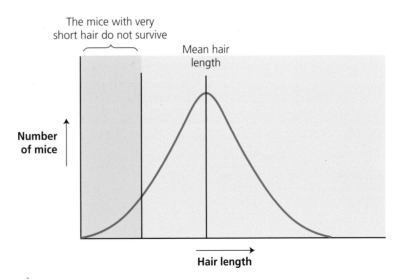

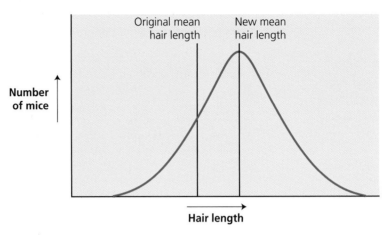

Figure 6.8 Directional selection

House mice are very adaptable species and are found in many different environments. Some live in cold stores. Mice living in cold stores will have an advantage if they have long hair; they will have a disadvantage if they have short hair. Over a period of time we might expect selection to result in mice with longer hair.

Stabilising selection

This operates against both extremes. Birth mass in humans provides an example.

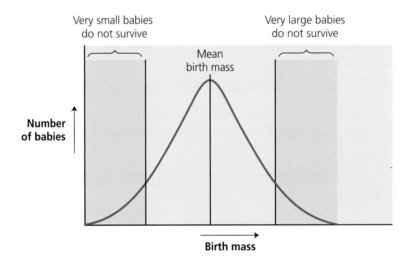

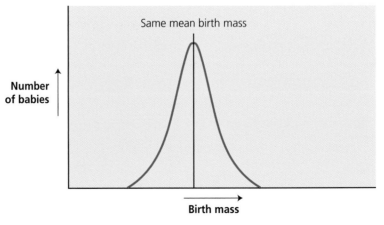

Figure 6.9 Stabilising selection

The *x*-axis of the graph in Figure 6.9 shows the mass at birth of a large number of human babies. Smaller babies are often underdeveloped and have a lower rate of survival; very large babies are also less likely to survive. Selection then, favours those in the middle so there is a tendency towards less variation. This is stabilising selection.

Disruptive selection

It is difficult to find a simple example with which to illustrate this type of selection, but the general idea is straightforward. Look at Figure 6.10.

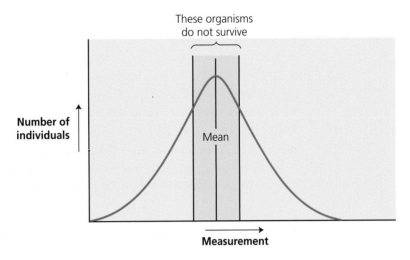

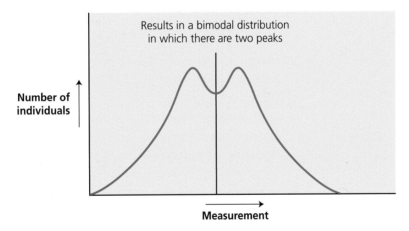

Figure 6.10 Disruptive selection

In this case, those organisms that are in the middle of the range are at a disadvantage and are selected against. The extremes will be favoured, and two different populations will gradually evolve.

6.3 Speciation

A species is defined as a group of organisms that can reproduce and produce fertile young. Speciation refers to the formation of a new species of organisms. **Intraspecific speciation** occurs when a single species gives rise to one or more new species. This requires isolating mechanisms to prevent gene flow between populations of the parent species. Natural selection may then act upon the isolated populations, leading to the development of new species.

Individuals may be prevented from reproducing due to a physical barrier preventing interaction, such as a mountain range or river. This is known as **geographical isolation** and may lead to **allopatric speciation**.

- Conditions on either side of the physical barrier will be slightly different.
- Organisms have to adapt to the different conditions — the natural selection processes will differ in each isolated group.
- Selection, together with the accumulation of different mutations in the two groups over a long period of time, will lead to divergence of the gene pools.
- Eventually, individuals from different populations will have changed so much that they will not be able to breed with one another to produce fertile offspring — they have become two separate species.

The finches of the Galapagos Islands are a good example of allopatric speciation. Although the finches had a common ancestor, geographical isolation has resulted in the formation of fourteen different species of finch. Each species of finch inhabits a different ecological niche on the islands and are adapted to local environmental conditions.

Reproduction may be prevented by various mechanisms other than geographical isolation. These mechanisms are associated with mating, fertilisation and the viability of offspring, known collectively as **reproductive isolation**. There are several causes of reproductive isolation.

- **Seasonal isolation** — individuals of the same species have different flowering or mating seasons.
- **Mechanical isolation** — mutations cause changes in the structure of genitalia, preventing successful mating.
- **Behavioural isolation** — individuals may be active at different times of the day, or develop courtship rituals that are not attractive to the main population of a species.
- **Gametic isolation** — male and female gametes fail to fuse, or to produce a viable embryo.

Speciation due to reproductive isolation is known as **sympatric speciation**.

Most speciation is **intraspecific** — new species arise from a common ancestor due to allopatric or sympatric speciation, or a combination of the two processes. **Interspecific** speciation (a new species arises due to the combination of two parent species) can also occur, but is rare in animals. It can be seen in plants as a result of **allopolyploidy**.

1 *Nervous communication*

When you have finished revising this topic, you should:

- understand that nervous communication involves detection of stimuli by receptors, transmission of nerve impulses, and responses by effectors
- be able to describe the structure of a neurone
- be able to explain how nerve impulses are initiated and propagated
- be able to describe transmission across a synapse

1.1 The essentials of nervous communication

Stimulus

A stimulus is a specific feature of an organism's environment that can be detected in some way. Stimuli may be features of the external environment, such as sound, pressure or light; or they may be features of the internal environment, such as blood temperature or water potential. Stimuli are the signals to which an organism responds.

Receptor

A receptor is a cell or an organ that is able to detect a particular stimulus which indicates a change in the environment of an organism. Receptors have a number of characteristics:

- They are specific; a particular receptor will only respond to one kind of stimulus. One kind of taste receptor on the tongue, for example, only responds to bitter-tasting substances. Substances that taste sour or sweet will not produce a response.
- They are extremely sensitive and can detect very small changes in the environment.
- They convert sensory information into a form that may be transmitted through the nervous system.

Response

This is an action in a muscle or a gland which arises as a result of stimulation by a nerve. Other responses result from chemical stimulation by hormones and plant growth substances.

Effector

This is an organ which responds to stimulation by the nervous system and produces a particular response. Muscles and glands are effectors and respond to stimulation by either contraction or secretion.

1.2 Neurones

A **neurone** is a nerve cell. It is a very specialised cell and its main features are:

- It has a large **cell body** containing the nucleus and many of the other cell organelles.
- Coming from this cell body are numerous fine branches called **dendrites**. These form connections or **synapses** with other nerve cells.

Nerves transmit
impulses, not messages.

- There are one or more larger branches called **axons**. An axon carries a nerve impulse in a direction away from the cell body. In mammals, many axons are surrounded by a sheath of fatty material known as **myelin**. The myelin enables the neurone to conduct nerve impulses rapidly.

A neurone which carries sensory information from a receptor is called a **sensory neurone**. A **motor neurone** is a neurone that carries an impulse to an effector.

1.3 Nervous transmission

A nerve impulse involves changes in electrical potential across the plasma membrane as the impulse passes along a nerve cell. To understand what causes these changes, look at each of the numbered features on the graph in Figure 7.1.

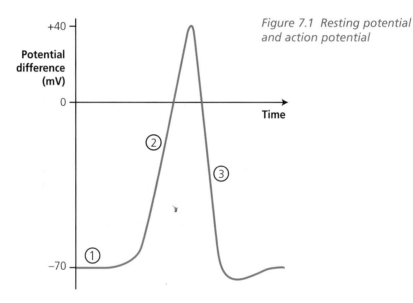

Figure 7.1 Resting potential and action potential

Make sure that you do not confuse the distribution and part played by sodium and potassium ions in the conduction of a nerve impulse.

1 When a nerve cell is at rest, there is a potential difference across the cell surface membrane. There is a higher concentration of potassium (K^+) ions and a lower concentration of sodium (Na^+) ions in the cytoplasm of the cell than outside. The membrane allows the potassium ions to diffuse out readily but it is much more difficult for the sodium ions to diffuse in. As a result, there are fewer positive ions inside the cell than outside. In other words, the inside of the nerve cell has a potential difference of approximately −60 mV when compared with the outside. This is called the **resting potential** of the cell.

   The plasma membrane of a nerve cell contains proteins which act as ion channels. Some of these channels allow sodium ions to pass through into the cell; some allow potassium ions to diffuse out of the cell into the cytoplasm. When the cell is at rest, these channel proteins are shut.

2 When a nerve impulse is initiated, a slight change in the potential difference across the membrane causes the sodium channel proteins to open and sodium ions to enter the cell. Since these ions are positive, they will make the inside of the nerve cell less negative. We say that the membrane is **depolarised** because the potential difference across the membrane has changed. As more sodium ions enter, the membrane becomes more depolarised and yet more sodium ions move into the cell. Eventually, the

inside of the cell becomes positive and the potential difference across the membrane rises to a value of approximately +40 mV. The sodium channel proteins now close.

3 The potassium channel proteins open much more slowly. They allow potassium ions to leave the cell. As potassium ions are positive, and they are leaving the cell, the potential difference becomes negative again and the resting potential is restored. The changes in potential difference described above form an **action potential**.

An active transport mechanism in the plasma membrane pumps out the sodium ions that have entered the cell during the action potential. They also pump the potassium ions back in.

1.4 Thresholds and all or nothing signals

When the nerve cell is stimulated, the initial stimulus must depolarise the plasma membrane enough to open some of the sodium channel proteins. If the stimulus is too small, this will not happen and an action potential will not result. The **threshold value** is the level that must be overcome in order to trigger an action potential.

Once the threshold value is overcome, however, this whole chain of events is set in motion and will always produce the same change in the potential difference across the membrane. For this reason, it is referred to as an **all or nothing** signal. A neurone can only give information about stimuli of different intensity by varying the frequency of nerve impulses; it cannot vary the size of the action potential.

1.5 Nerve impulses

A **nerve impulse** results from the spread of an action potential along a neurone. The change in potential difference that occurs at the point where the neurone is stimulated causes the next bit of the plasma membrane to become depolarised. In this way, a wave of depolarisation spreads along the neurone. In a neurone that is not surrounded by a myelin sheath, the speed of conduction is approximately 2 m s^{-1}. In neurones with myelin sheathes, the nerve impulse travels much faster. This is because depolarisation can only take place where there are gaps in the myelin. As a result, conduction is described as **saltatory** because it goes in a series of rapid jumps from one gap to the next.

1.6 Synapses and synaptic transmission

Nerve impulses must pass from one nerve cell to the next in order to provide the necessary signals between receptors and effectors. Nerve cells do not join directly to each other; they are separated by small gaps called synapses. Figure 7.2 is a simplified diagram showing one particular type of synapse.

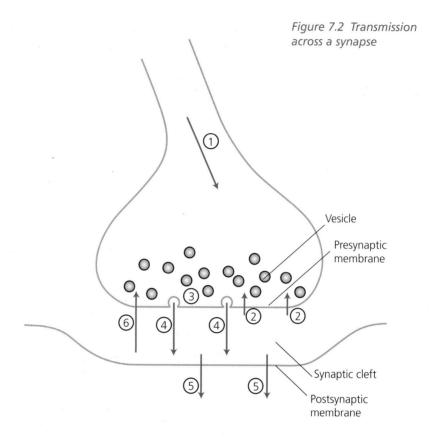

Figure 7.2 Transmission across a synapse

The stages in transmission across a synapse are:

1 A nerve impulse travels down the axon of the presynaptic neurone and arrives at the **presynaptic membrane**.

2 Calcium channel proteins in the presynaptic membrane open and allow calcium ions to diffuse into the neurone.

3 As a result of the increase in calcium ion concentration, some of the **vesicles** fuse with the presynaptic membrane and burst, releasing the **neurotransmitter, acetylcholine**, into the synaptic cleft.

4 The acetylcholine diffuses across the synaptic cleft and binds to receptors on the **postsynaptic membrane**.

5 Sodium channel proteins open and allow sodium ions to enter the postsynaptic neurone. This produces an action potential in the postsynaptic neurone.

6 The acetylcholine is almost immediately hydrolysed by the enzyme **acetylcholinesterase** into its components. These are reabsorbed through the presynaptic membrane and used to produce more acetylcholine.

The sequence of events described above occurs in a **cholinergic excitatory synapse** — cholinergic because the neurotransmitter is acetylcholine, and excitatory because transmission across the synapse gives rise to an action potential in the postsynaptic neurone. There are, however, other types of synapse that work in slightly different ways. There are neurotransmitters besides acetylcholine. They include **noradrenaline**, which is found in synapses in the sympathetic nervous system, and **serotonin**, which is found in the brain.

Some synapses are **inhibitory** and make it less likely that an action potential will arise in the postsynaptic membrane. Whether or not a particular synapse is excitatory or inhibitory depends on the receptor sites on the postsynaptic

membrane. If it is inhibitory, it affects chloride and potassium channel proteins rather than those allowing sodium ions through the membrane.

2 *Homeostasis*

When you have finished revising this topic, you should:

- understand what is meant by homeostasis and be able to explain why it is important in living organisms
- understand what is meant by negative feedback and be able to explain how it is involved in homeostasis
- be able to explain how temperature is controlled in a mammal
- be able to explain how blood glucose concentration is controlled in a mammal

2.1 Why homeostasis is important

Homeostasis involves keeping conditions inside an organism constant. There are a number of reasons why homeostasis is particularly important in living organisms:

- Biochemical reactions are controlled by enzymes. Changes in pH and temperature affect the rate of enzyme-controlled reactions. In extreme cases this can lead to the denaturing of enzymes and other proteins.
- External factors such as temperature often fluctuate considerably. A constant internal environment allows independence from these fluctuations. This enables animals such as mammals to live in conditions ranging from arctic to tropical.
- Water moves in and out of cells by osmosis. By maintaining a constant water potential in the surrounding tissue fluid, osmotic problems are avoided.

2.2 Negative feedback and homeostasis

Features of an organism's internal environment such as temperature, pH and the concentration of many dissolved substances have a set level. **Negative feedback** is the process in which a fluctuation from this set level sets in motion changes which return it to its original value. In humans, body temperature is about 37°C. Vigorous exercise will cause the temperature to rise. This produces a number of changes such as increased sweating that result in loss of heat and the body temperature falling again. Negative feedback is summarised in Figure 7.3.

Look carefully at the arrangement of this diagram. We will use it wherever we need to show how negative feedback operates. It is much easier to learn something if you can see a common pattern.

Examiners frequently ask candidates to explain what is meant by negative feedback. Make sure you know what this term means.

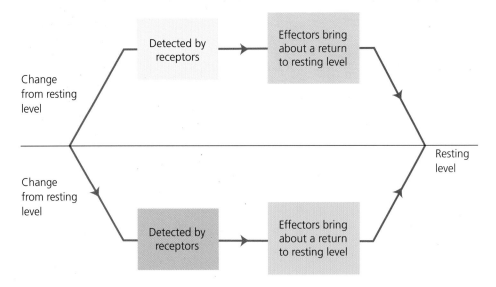

Figure 7.3 Negative feedback

2.3 Temperature control

In order to function normally, animals must keep their temperatures within certain limits. If an animal's temperature falls too low, biochemical reactions will be too slow for it to remain active. If it goes too high, there is a risk of enzymes and other proteins being denatured. To a certain extent, all animals exert some control over their internal temperatures. Mammals rely largely on physiological methods. Humans, for example, maintain a temperature within a degree or so of 37°C. If the temperature rises too high, various internal mechanisms bring about loss of heat. If it falls too low, other mechanisms serve to increase heat production and reduce its loss from the body. Humans are, therefore, examples of **endothermic** animals. Crocodiles, on the other hand, rely largely on moving between land and water to remain in the environment with the most suitable temperature. They are referred to as **ectothermic**, as they rely on the external environment to maintain a reasonably constant internal environment.

Temperature control in a mammal is achieved by negative feedback. If the temperature rises, the resulting increase in blood temperature is detected by receptors in the hypothalamus of the brain. As a result, the **heat loss centre** which is also in the hypothalamus sends nerve impulses to structures such as arterioles and sweat glands in the skin which bring about the necessary fall in temperature. Cold conditions, on the other hand, are detected by receptors in the skin. Impulses are sent to the hypothalamus. This time, the **heat conservation centre** triggers mechanisms which conserve the body's heat or even generate more heat by actions such as shivering. The outcome, once again, is a return to the normal level. This is summarised in Figure 7.4.

As body temperature increases, arterioles dilate and more blood flows through the capillaries which lie close to the skin. The very thin walls of the capillaries do not contain muscle, so they cannot dilate or constrict.

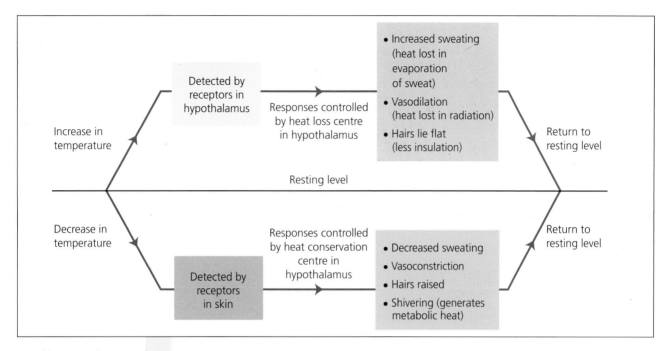

Figure 7.4 Controlling blood temperature

Vasodilation results in more blood flowing nearer to the surface of the body. It does not involve blood vessels moving up and down in the skin.

2.4 Control of blood glucose

Accurate control of blood glucose concentration is very important. If the concentration falls too low, the central nervous system does not function correctly. If it rises too high, there will be a loss of glucose from the body in the urine. Although the overall concentration stays within narrow limits, it is important to appreciate that glucose is always being added to and removed from the pool of glucose in the blood. Digestion and the absorption of carbohydrates and conversion from stores of glycogen and fat, tend to increase blood glucose concentration, while processes such as respiration decrease it.

Insulin is a hormone produced by the **islets of Langerhans**, groups of cells in the pancreas. The secretion of insulin is stimulated by the rise in blood glucose concentration which follows a meal. The hormone has a number of effects on the body, all of which tend to lead to a reduction in the concentration of glucose in the blood. Two of these effects are:

1 Insulin speeds up the rate at which glucose is taken into cells from the blood. Glucose normally enters cells by facilitated diffusion through protein carrier molecules in the plasma membrane. Cells have extra carrier molecules present in their cytoplasm. Insulin causes these carrier molecules to be sent to the membrane where they increase the rate of glucose uptake by the cell.

2 It activates enzymes which are responsible for the conversion of glucose to glycogen.

Glucagon and glycogen — very similar words, very different meanings. Spell them correctly!

Glucagon is another hormone produced by the islets of Langerhans. The secretion of glucagon is stimulated by a fall in blood glucose concentration. The main effect of this hormone on the body is to activate enzymes in the liver which are responsible for the conversion of glycogen to glucose. It also stimulates the formation of glucose from other molecules such as amino acids. The control of blood glucose concentration is summarised in Figure 7.5.

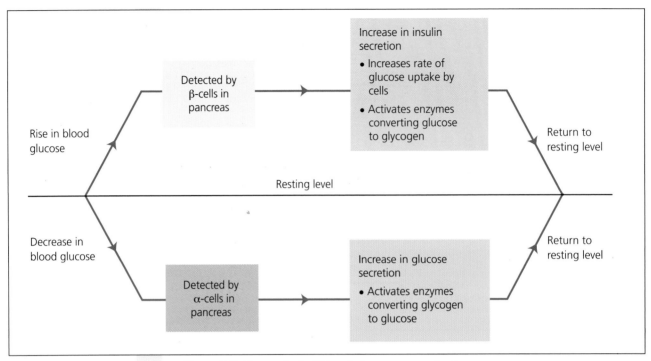

Figure 7.5 The control of blood glucose concentration

3 ## Excretion and the kidney

When you have finished revising this topic, you should:

■ be able to explain what is meant by excretion and describe how urea is formed in the liver

■ be able to draw a simple diagram showing the main features of a nephron

■ be able to explain how the formation of urine involves:

– ultrafiltration in the renal capsule

– selective reabsorption in the first convoluted tubule

– concentration in the loop of Henle and the collecting duct

■ be able to explain the part played by the kidney in controlling the water balance of the body

3.1 Excretion

Metabolic reactions produce waste products. Excretion is the removal of these waste products from the body. If more protein is eaten than is required, the excess amino acids cannot be stored. The amino group is removed (**deamination**) and converted into ammonia. Ammonia is very toxic and can only be excreted safely in animals that live in water, where it can be diluted to safe concentrations. In mammals, ammonia is converted to a less poisonous compound, urea, in a cycle of biochemical reactions called the **ornithine cycle**. The formation of urea is summarised in Figure 7.6.

Distinguish carefully between urea and urine.

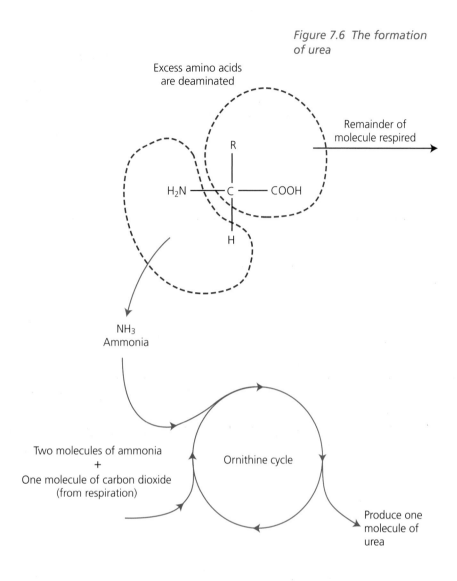

Figure 7.6 The formation of urea

Excess amino acids are deaminated

Remainder of molecule respired

NH_3
Ammonia

Two molecules of ammonia
+
One molecule of carbon dioxide
(from respiration)

Ornithine cycle

Produce one molecule of urea

3.2 Nephrons and urine formation

Urine is formed in the kidney in the thousands of nephrons. Each **nephron** is a long tubule which is divided into a number of distinct regions. The first part of the nephron is the **renal capsule**, and it is here that the process of **ultrafiltration** takes place. This process separates the smaller, soluble substances from the blood. The filtrate contains excretory products and substances useful to the body. As it passes down the **first convoluted tubule**, its composition is changed as a result of selective reabsorption. Useful substances such as glucose and amino acids are reabsorbed into the blood. Finally, the **loop of Henle** and the **collecting duct** play an important part in reabsorbing water and producing concentrated urine. The process of urine formation is summarised in Figure 7.7.

Ultrafiltration

Each renal capsule consists of a ball of capillaries called a **glomerulus** surrounded by the funnel-like structure of the renal capsule itself. The capsule consists of an outer layer of cells, separated by a space from an inner layer which is wrapped closely round the glomerulus. The basement membrane of the cells lining the capillary walls acts as a filter. The pressure of the blood in the glomerulus forces fluid through the basement membrane and into the capsule space. This fluid has

Get an overall view of what happens in a nephron before you start to concentrate on learning the details.

the same composition as blood plasma except that it does not contain protein. Protein molecules are too large to pass through.

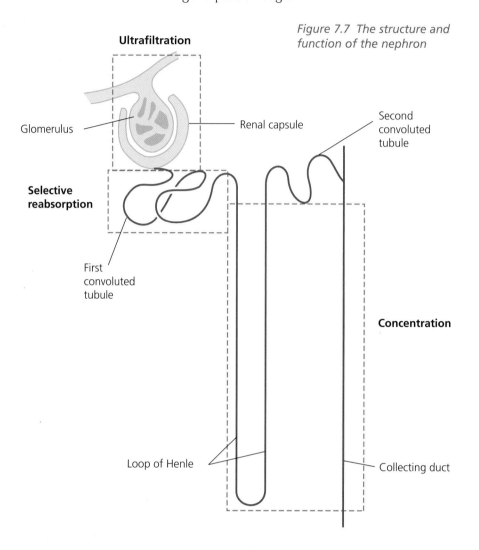

Figure 7.7 The structure and function of the nephron

First and second convoluted tubules might be easier to learn and less easy to confuse than proximal and distal convoluted tubules.

The cells lining the first convoluted tubule possess microvilli, not villi.

Selective reabsorption

In the first (proximal) convoluted tubule:

- Useful substances such as glucose, amino acids and some mineral ions are reabsorbed. This involves active transport. The cells of this part of the tubule show a number of adaptations that can be linked with their function. They have microvilli that increase their surface area and they contain large numbers of mitochondria that provide the ATP necessary for active transport.
- The removal of soluble substances means that a water potential gradient is established. Water will therefore be reabsorbed from the nephron by osmosis.
- Excretory products such as urea are not reabsorbed. Because water has been removed, however, the concentration of urea in the first convoluted tubule becomes more and more concentrated as the fluid flows along the tubule.

Producing concentrated urine

To understand how the loop of Henle and the collecting duct produce urine that is more concentrated than blood plasma, you will need to look at Figure 7.8.

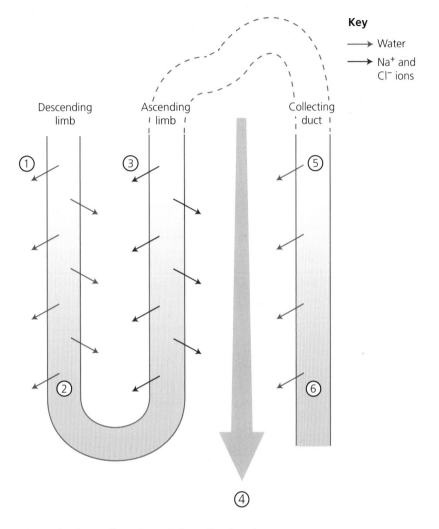

Figure 7.8 The loop of Henle and the collecting duct

Simplifying the overall picture as much as possible, what happens is that:

1 The **descending limb** of the loop of Henle is permeable to water but is impermeable to sodium and chloride ions. Since there is a higher concentration of dissolved substances in the surrounding kidney tissue, water will move out of the nephron by osmosis down the resulting water potential gradient.

2 Loss of water in this way results in the ion concentration increasing as the fluid passes down the descending limb of the loop of Henle.

3 The **ascending limb** of the loop of Henle is permeable to sodium and chloride ions but impermeable to water. These ions are pumped from the ascending limb by active transport.

4 The result of this is that from the outside of the kidney to the inside there is an increasing concentration gradient in the tissue surrounding the tubule.

5 The **collecting duct** is permeable to water. Since the concentration of ions is greater in the surrounding tissue than in the collecting duct, water will move out of the collecting duct by osmosis.

6 The concentration gradient in the kidney tissue ensures that there is always a water potential gradient between the collecting duct and the surrounding tissue. Water will therefore continue to be removed, and this will result in concentrated urine.

The longer the loop of Henle, the greater the concentration gradient that can be established in the kidney tissue and the more concentrated the urine. This explains why small mammals that live in deserts have extremely long loops of Henle.

3.3 The kidney and water balance

In very hot conditions, large amounts of water are lost as sweat. If there is insufficient drinking water to make up for this loss, the water content of the blood will fall. This has extremely serious consequences. Not only will the osmotic balance of the body be upset but the blood will become thicker or more viscous and this could lead to circulatory failure.

Antidiuretic hormone (ADH) plays an important role in conserving water in such circumstances. It is secreted by the pituitary gland and acts by increasing the amount of water that is reabsorbed in the collecting duct. A negative feedback mechanism is involved.

A negative feedback mechanism is involved in control of water balance, so we can use a similar diagram in Figure 7.9 to the ones we used to illustrate the control of concentration of body temperature and blood glucose concentration in the previous section.

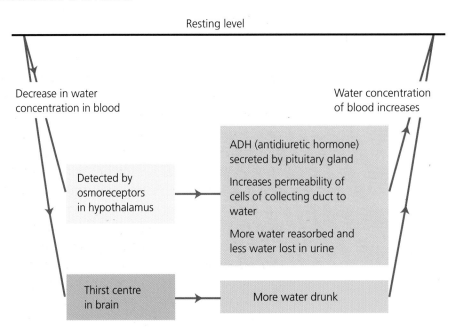

Figure 7.9 ADH and the control of water balance

It is worth noting that we only have a specific mechanism to conserve water when it is in short supply in the body. If the water concentration in the blood is higher than normal, secretion of ADH is inhibited. The large amounts of dilute urine produced as a result rapidly return the water concentration to its set level.

CHAPTER 8 Ecosystems

1 Ecology: what it's all about?

When you have finished revising this topic, you should:

- be aware of most of the common problems that candidates encounter in producing answers to ecology questions
- understand the meanings of the basic terms used in writing about ecological topics

1.1 Revising ecology

For most A-level biology students, ecology causes problems. Before you start your revision, it might be worth looking at this topic from the point of view of the examiner. Why is it that so many students fail to gain marks on questions concerned with ecology? In this first section we will look at three main areas of concern and suggest some steps you might take to avoid making the same errors. At the very least, being aware of the problem should help.

Background knowledge

Examiners have to stick to the specification. They can't ask you for facts about subjects which are not spelt out in the specification but they may expect you to use your knowledge to interpret information which is unfamiliar. For example, competition is probably part of your specification although potatoes are probably not. It would not be fair to set a question which asks for lots of detail about growing potatoes but it would be perfectly reasonable to ask you to interpret some information about competition between potato plants and weeds. By the time you have opened this book and decided to do some serious revision, it is probably getting a bit late to start thinking about your background knowledge. But there are one or to things that you might do:

- There are some outstanding natural history programmes on the television. Find time to relax a bit and watch two or three of them. Try to find programmes which relate to this country if possible.
- Find a couple of articles from magazines, *Biological Sciences Review* or *BBC Wildlife* for example, and read through them.

Recognising that different species of organisms differ in the food they eat and the way they live

This point is probably best explained with an actual example. Suppose a question asks about food chains and you decide to illustrate what you have written with this example:

$$\text{grass} \rightarrow \text{insect} \rightarrow \text{sparrow} \rightarrow \text{hawk}$$

This seems fine, but let us have a critical look at it. Take the first link. There are more species of insect than of any other group of living organisms. Some, like the caterpillars of certain moths and many grasshoppers will eat grass, but a lot of them won't. The term 'insect' then is much too vague. How about the next link. Sparrows are seed eaters; they do not feed on insects — another mistake.

Throughout this section on ecology we will use examples to illustrate general ideas. The organisms we will use are shown in Figure 8.1. Many of these should

Other animals don't behave and think like humans! Let's consider the two-spot ladybird. If there are no aphids for it to eat, it will die. It certainly won't turn vegetarian and feed on something else. Remember, animals have specialised feeding habits and these limit what they can and will eat.

be familiar to you already, and they can all be found living in a clump of nettle plants.

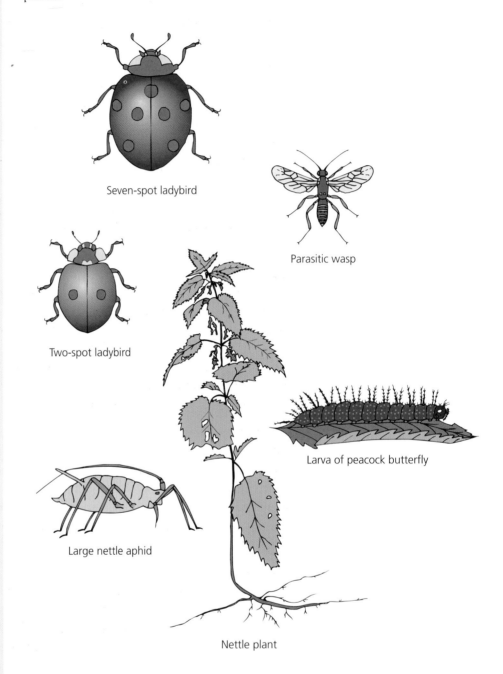

Figure 8.1 Some of the common organisms which can be found on a clump of nettle plants

Use of appropriate A-level language

It is not only what you say but how you write it that is important. Animals and plants **compete** for resources; they don't 'fight' for them. Organisms have specific ecological **niches** in particular **habitats**, they don't live in 'nests' or have 'homes'. It is much easier to do something about this.

Here is a list of eight basic ecological terms. Get to know them, get to know what each one means and, above all, get to use them.

Environment

The environment of an organism is the set of conditions which surround it. The environment consists of a non-living or abiotic component and a living or **biotic** component.

Abiotic

Abiotic factors are those which relate to the non-living part of the environment. Nettle plants, for example, grow particularly well where there are high concentrations of phosphate ions in the soil. On the other hand, the number of aphids found on a nettle plant will be influenced by temperature and rainfall. Mineral ion concentration, temperature and rainfall are all abiotic factors.

Biotic

Biotic factors are those which relate to the living part of the environment. Competition between seven-spot ladybirds and two-spot ladybirds for aphids and the effect of predation on the numbers of aphids are examples of biotic factors.

Population

A population is a group of individuals belonging to a particular species. The members of a particular population will be found in a particular place at the same time. The two-spot ladybirds living in early spring on a patch of nettles growing in a corner of a field will form a population. Those that live on another patch of nettles on the other side of the field will form another population. Members of the same population are generally able to breed with one another; members of different populations seldom do.

Community

A community is the term used to describe all the populations of different organisms living in a particular place at a particular time. If we go back to our example of a patch of nettles, then the community is made up of all the living organisms present: the nettles, the aphids, the ladybirds and the various other organisms that we have not mentioned, such as the fungi and bacteria which live in the soil.

Ecosystem

An ecosystem forms the basic unit of ecology. It consists of the community of living organisms and the abiotic factors which influence them. Although ecosystems are generally considered as being separate from one another, they do influence each other. The use of pesticides on farmland in Europe and North America, for example, has resulted in an increased pesticide concentration in the tissues of the penguins that live in Antarctica.

Habitat

The habitat is the place where a particular population or community lives.

Niche

The meaning of this term is rather complicated. In very simple terms, the niche of an organism is the place where it is found and what it does there. It is a description of how an organism fits into its environment. Let us try to explain this by looking at the example of a two-spot ladybird. We can describe the niche of this insect in terms of the features of the abiotic environment in which it lives, such as the temperature range it can tolerate and the position on the plant where it feeds. Our description should also describe its feeding habits, referring to the size and species of aphids on which it feeds.

2 *Nutrient cycles*

When you have finished revising this topic, you should:

- understand the general principles involved in the cycling of nutrients
- know the basic details of the carbon cycle
- be able to describe the part played by microorganisms in the cycling of carbon
- understand how the cycling of carbon is similar to the cycling of other nutrients
- know how human activities influence the cycling of carbon
- know the basic details of the nitrogen cycle
- be able to describe the part played by microorganisms in the cycling of nitrogen
- understand how the cycling of nitrogen is similar to the cycling of other nutrients
- know how human activities influence the cycling of nitrogen

2.1 Introduction

The most important thing about the ways in which elements such as carbon and nitrogen are cycled is that the processes involved are very similar. The key thing to understand is the basic cycle shown in Figure 8.2.

> All you need to do is to learn this. It can be adapted to represent the carbon cycle, the nitrogen cycle or any other cycle.

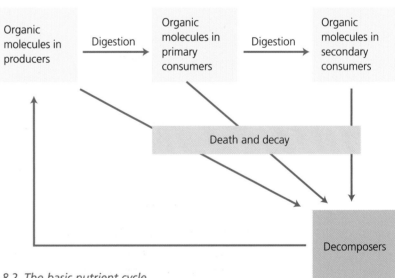

Figure 8.2 *The basic nutrient cycle*

Look at this cycle carefully. Let us start with the element identified as part of an organic molecule in a producer. For the carbon cycle, this is carbon in molecules such as those of carbohydrates, lipids or proteins; for the nitrogen cycle, it is nitrogen contained in protein molecules.

The producer may be eaten by a primary consumer. If this is the case, the organic molecules concerned are digested. The simple, soluble molecules produced are absorbed and built up in the body of the primary consumer, a process known as **assimilation**. The same thing happens when the primary consumer is eaten by a secondary consumer. The first step, then, is the passage of the element concerned along the food chain from producer to primary consumer to secondary consumer.

The next step occurs when one of these organisms dies. Its tissues are decomposed by bacteria and other microorganisms. Ultimately, its organic molecules are broken down and the element is released as part of an inorganic molecule or ion. This can then be taken up again by producers. This is the basic cycle. The carbon and nitrogen cycles, like all nutrient cycles, are based on this.

2.2 The carbon cycle

Most textbooks tend to emphasise the differences between different cycles. It is much easier to understand what is going on if you concentrate on the ways in which they are similar. The best way to learn how carbon is cycled in an ecosystem is to start from the basic nutrient cycle in Figure 8.2. Learn it and copy it out. If you modify this to show a carbon cycle it will probably look something like Figure 8.3.

This cycle has been modified from the basic nutrient cycle shown in Figure 8.2.

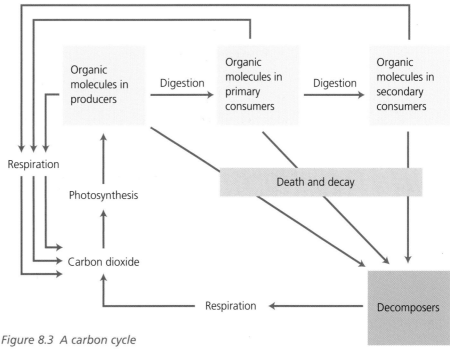

Figure 8.3 A carbon cycle

The particular points that you need to note are:

Plants take up carbon in the form of carbon dioxide during the process of photosynthesis. They do not take up carbon-containing ions through their roots.

- The only way in which carbon can be taken up by producers is as carbon dioxide. This is absorbed by the leaves of a plant and converted into organic compounds in the process of photosynthesis.
- Carbon dioxide is released as a result of respiration by bacteria, by producers, and by primary and secondary consumers.
- **Saprophytes**, or as they are more correctly called, **saprobionts**, are organisms that obtain their nutrients from dead or decaying organic matter. Many

bacteria are saprobionts. They secrete enzymes which digest the complex carbon-containing molecules in dead organic matter. The smaller molecules which result are absorbed and some of them are used in respiration, releasing carbon dioxide.

To sum up, all you need to do is to learn and be able to draw the basic nutrient cycle. Then alter it to represent a carbon cycle. Don't worry about whether it looks exactly the same as the one shown here and don't waste time with rulers, colour and pictures of bonfires and dead mice!

2.3 Human activities and the carbon cycle

The carbon cycle that we have looked at so far has ignored the fact that large amounts of carbon are locked up in oil, coal and limestone that were formed millions of years ago. There are also large amounts of carbon contained in forests, some of which have existed for many thousands of years. The ways in which human activities influence the carbon cycle are nearly all involved with releasing this 'stored' carbon as carbon dioxide. They include:

- **Combustion of fossil fuels, such as coal and oil.**
- **Clearing of forests.** It is important to appreciate that living rain forests have little effect on the amount of carbon dioxide in the atmosphere. This is because the amount of carbon dioxide they remove in the process of photosynthesis is about the same as that added by the respiration of all the organisms that live there. However, when a forest is cleared for agriculture, the carbon locked up in the trees is released as carbon dioxide when they are burnt.
- **Weathering of limestone.** Limestone is a rock formed from the skeletons of tiny marine organisms. It consists mainly of calcium carbonate. Using limestone in buildings increases the surface area exposed to the normal weathering processes which produce carbon dioxide.

2.4 The nitrogen cycle

Any examiner will tell you that the nitrogen cycle creates more difficulties for examination candidates than almost any other single topic in the A-level biology specification. Despite the fact that you can almost guarantee that there will be a question on the nitrogen cycle, it always comes as a surprise that so few students appear to understand it. So, there is a message here. Make sure you really know your nitrogen cycle.

Let's start by looking at the basic cycle. We can use exactly the same approach as we did with the carbon cycle. Start from the general nutrient cycle in Figure 8.2. Now modify it. At this stage we will not consider the processes of denitrification and nitrogen fixation. We will look simply at the way in which nitrogen in the organic molecules of a dead rat, for example, is made available to producers in the form of nitrate. This is shown in Figure 8.4.

There are a number of nitrogen-containing substances in a plant, including amino acids, proteins and nucleic acids. This nitrogen is passed from producer to primary consumer to secondary consumer in a series of stages which involve digestion and assimilation. When organisms die, the nitrogen-containing organic matter is digested by saprobiotic bacteria and ammonia is released.

Another group of bacteria, the **nitrifying bacteria**, then convert ammonia to nitrites and nitrites to nitrates. Nitrates are taken up from the soil by producers, and the cycle is complete.

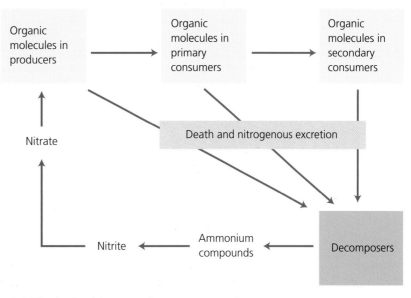

Figure 8.4 The basic nitrogen cycle

Questions often ask you to describe the pathways by which nitrogen in one form is converted to nitrogen in another. You don't need to incorporate denitrification and nitrogen fixation every time.

The particular ways in which this part of the nitrogen cycle differs from the general pattern are:

- Nitrogen from consumers is made available to saprobiotic bacteria when they die and through excretory products such as urea.
- Two steps are involved in converting organic nitrogen-containing molecules into the inorganic ions which can be taken up by the producers:
  1 Conversion of organic nitrogen-containing molecules such as proteins to ammonia by saprobiotic bacteria.
  2 Conversion of ammonia to nitrites and nitrites to nitrates. This is done by nitrifying bacteria.

This is the basic cycle; but there is a complicating factor. Under anaerobic conditions such as those that exist when the soil is water-logged, some bacteria are able to get the oxygen they need from nitrates. These are **denitrifying bacteria**. Effectively, what they do is to convert nitrates to nitrogen which escapes into the atmosphere. Unfortunately, nitrogen itself cannot be used by animals and plants, so **denitrification** removes nitrogen from the cycle.

Use the word 'nitrogen' only when you mean the chemical element. Otherwise use terms such as nitrates or nitrogen-containing compounds. This will help you to avoid confusion.

However, atmospheric nitrogen can be made available to plants by the process of **nitrogen fixation**. Nitrogen fixation is the incorporation of nitrogen gas into organic nitrogen-containing compounds. There are several ways in which nitrogen can be fixed, and all of them require a considerable amount of energy:

- **Lightning.** Electrical energy in lightning combines nitrogen and oxygen in the air to produce various oxides of nitrogen. Rain washes these into the soil where they are taken up by plants as nitrates.
- **Ammonium fertilisers.** These are made by an industrial process which combines nitrogen directly with hydrogen.

- **Nitrogen-fixing microorganisms living in the soil.** Some soil bacteria produce the enzyme nitrogenase. These bacteria are able to use this enzyme together with energy from ATP to convert nitrogen to ammonium compounds and then to organic nitrogen-containing substances.
- **Nitrogen-fixing microorganisms living in the root nodules of leguminous plants.** This is similar to the previous process. The bacteria convert nitrogen into ammonium compounds, some of which pass to the plant. The bacteria gain carbohydrates from the plant.

Let us return to our diagram of the nitrogen cycle. It is better to add this information about denitrification and nitrogen fixation on as a sort of 'bolt-on' extra rather than end up with a very complicated diagram which you will have difficulty remembering. Figure 8.5 is a simple diagram of the full nitrogen cycle and incorporates this extra material.

> Nitrogen-fixing bacteria convert nitrogen into ammonium compounds which in turn become incorporated into organic compounds. They do not convert nitrogen into nitrates.

> It is better to add a note explaining the ways in which nitrogen fixation occurs rather than try to put everything in the diagram itself.

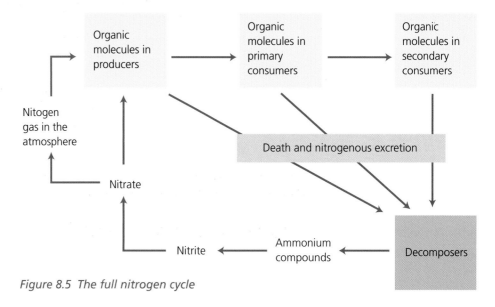

Figure 8.5 The full nitrogen cycle

To sum up, draw the general nutrient cycle and alter it to represent a basic nitrogen cycle. Add on the extra pathway involving denitrification and nitrogen fixation.

2.5 Human activities and the nitrogen cycle

There is a lot of nitrogen contained in chemical compounds in a crop of plants growing in a field. Harvesting removes this nitrogen. In order to grow another crop, fertiliser is added to replace the nitrates and other nutrients which the growing crop removed from the soil. Nitrogen-containing ions are very soluble and can be washed or **leached** from the soil to accumulate in lakes and rivers.

The breakdown of sewage and waste from cattle and pigs also results in the addition of nitrate to lakes and rivers.

The addition of large amounts of nutrients such as nitrates and phosphates to areas of fresh water results in eutrophication. This is summarised in the flow chart in Figure 8.6.

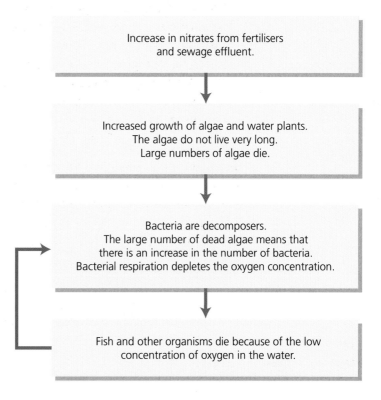

It is the respiration of decomposing bacteria which brings about the fall in oxygen concentration.

Figure 8.6 Flow chart showing how eutrophication may affect the organisms in a freshwater lake or river

3 *Food chains and food webs*

When you have finished revising this topic, you should:

■ be able to draw a basic food web for some of the organisms found on a nettle patch

■ be able to convert information in food chains and food webs into pyramids of number, biomass or energy

■ understand how energy is transferred through organisms in an ecosystem and is ultimately lost from the system as heat

■ know that the transfer of energy between trophic levels is inefficient and understand some of the factors which contribute to the efficiency of the transfer

3.1 A simple food web

As was pointed out at the start of this chapter, one of the examiners' most common criticisms of student answers on ecology and of those about food webs and energy transfer in particular, is that they are often extremely vague and inaccurate. We shall use the food web in Figure 8.7 to illustrate this topic. It shows some of the organisms living and feeding in a patch of nettles. These organisms are illustrated in Figure 8.1. There are several advantages in considering this particular food chain. The main one is that these are real organisms and most of them should be familiar to you so they will be easy to remember.

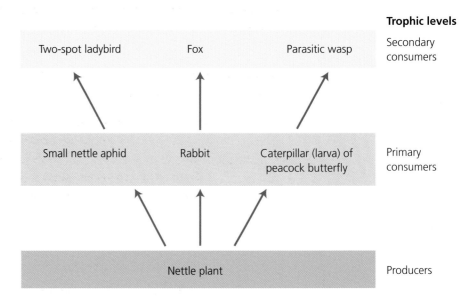

Note: The arrows represent the direction of energy flow.
The small nettle aphid therefore gets its energy from
nettle plants.

Figure 8.7 A food web showing some of the organisms living in a patch of nettles

In order to avoid making this web too complicated, the decomposers have been left out.

3.2 Ecological pyramids

Diagrams of food webs only show qualitative information. Pyramids of numbers, biomass and energy provide us with quantitative information. Figure 8.8 shows the appearance of these ecological pyramids for three food chains, all taken from the web in Figure 8.7.

If you are asked to draw an ecological pyramid, always label it. Use the terms producer, primary consumer and secondary consumer rather than trophic levels 1, 2, 3, etc.

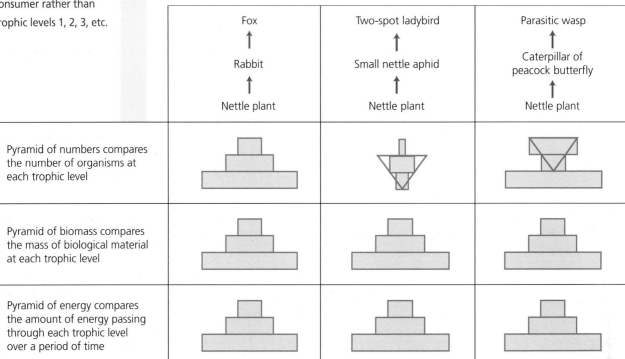

Figure 8.8 Ecological pyramids

There are points to note about each type of pyramid. All pyramids are pyramid-shaped other than the exceptions given below.

Pyramids of number

Pyramids of number allow us to compare the number of organisms present in each trophic level at a particular time.

Although pyramids of number are pyramid-shaped there are two important exceptions. Pyramids will be upside down or inverted if there are a lot of small animals feeding on a large plant. They are also inverted where an animal has a large number of small parasites feeding on it. One human, for example, can have a large number of head lice. In Figure 8.8, the butterfly larva is being fed on by a large number of parasitic wasps.

Pyramids of biomass

Biomass is a measure of the total amount of living material present.

Pyramids of biomass allow us to compare the mass of organisms present in each trophic level at a particular time. Pyramids based on biomass get over the problems of organisms differing in size.

Most pyramids of biomass are pyramid-shaped, but there is one important exception which you should know about. This is when the producer is a small organism which multiplies very rapidly. In this case, the total biomass of the producers present at any one time may be less than the total biomass of the primary consumers. The pyramid will again be inverted.

Pyramids of energy

Pyramids of energy allow us to compare the amount of energy passing through each trophic level over a period of time. They differ from the other two types of ecological pyramid which measure the number and biomass of organisms present in each trophic level at a particular time.

Pyramids of energy are always pyramid-shaped. There are no exceptions to this rule.

3.3 Energy transfer

The idea of the transfer of energy allows us to consider the efficiency with which light energy is transferred to energy in producers, as well as the efficiency with which energy in the producers is then transferred from trophic level to trophic level.

Look at Figure 8.9. This shows the percentage of energy transferred to each trophic level in the ecosystem. We can look at this another way. For every 10 000 kJ of energy absorbed by the producer, 100 kJ are incorporated into its tissues, 10 kJ will eventually be incorporated into the tissues of primary consumers, and 1 kJ into the tissues of secondary consumers. The rest will be lost as heat. This is the basic pattern of energy transfer, but there are a number of points that are worth making about each stage. These points are often required in order to answer questions which involve the interpretation of information.

Look carefully at the pyramids of energy. They show that there is more energy flowing through the primary consumers than through the secondary consumers. This does not mean that there is more energy in a primary consumer than in a secondary consumer.

You may well be asked to carry out simple calculations to show the percentage of energy transferred between trophic levels. Have a look at the questions produced by your examining board and practise these calculations.

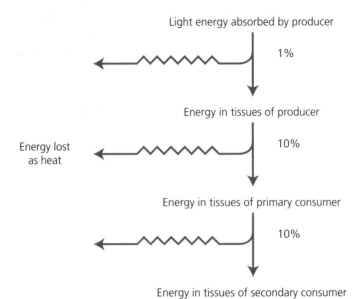

Light energy absorbed by producer

1%

Energy in tissues of producer

10%

Energy lost as heat

Energy in tissues of primary consumer

10%

Energy in tissues of secondary consumer

Figure 8.9 The efficiency with which energy is transferred within an ecosystem

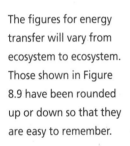

The figures for energy transfer will vary from ecosystem to ecosystem. Those shown in Figure 8.9 have been rounded up or down so that they are easy to remember.

Transfer of sunlight energy to energy in plant tissues

Not all the light energy falling on a plant is used to make new tissues:

- Some is of the wrong wavelength for photosynthesis.
- Some fails to strike a chlorophyll molecule.
- Some will be reflected from the plant surface.
- Other factors such as soil nutrients or carbon dioxide concentration may be in short supply. This will limit the rate of formation of new tissue.

Crop plants often convert a higher percentage of the light energy which falls on them into energy in new tissue than plants growing in the wild do. This is because:

- Crops are often irrigated and supplied with fertiliser. Shortage of water and mineral ions does not limit growth.
- Crop plants have been bred for high productivity. They therefore have genes which ensure that they are efficient at converting light energy into energy in plant tissue.
- Crops are often treated with pesticides. As a result, there is little damage to their leaves and they can photosynthesise more efficiently.

Transfer of energy from one trophic level to another

In Figure 8.9, the efficiency of energy transfer was given as 10%, but it is very variable. Some of the factors which result in this variation include the following:

- Carnivores are rather more efficient than herbivores at converting energy in food into energy in body tissues. This is partly because plants contain much more indigestible material than animals.
- **Poikilothermic** animals, such as fish, are much more efficient at transferring energy in food into energy in body tissues than **homoiothermic** animals, such as pigs and chickens. This is because a homoiothermic animal maintains a high body temperature. It does this by producing heat by metabolic processes. A lot of the food that a homoiotherm eats goes into heat production.

• A small mammal has a larger surface area compared with its volume than a large mammal has. It therefore loses more body heat. Since this heat comes from its metabolic processes, it follows that a lot more of the energy in the food that a small mammal eats is lost as heat and does not go into making new tissues.

4 *Counting and estimating*

When you have finished revising this topic, you should know how to:

■ use quadrats and transects to sample populations

■ obtain a random sample and understand why random sampling is important

■ estimate the number of animals in a particular area using the mark-release-recapture technique

4.1 Sampling

Ecologists often need to count the number of organisms present and find out about their distribution. There are very few species where it is possible to count all the individuals in a particular population. Normally they take **samples**. Whatever method of sampling is used, there are two important principles to bear in mind:

1 The sample must be large enough. The larger the size of the sample, the more accurate the results. A very small sample could differ from the rest of the population. This is a matter of chance. We must make sure that any sample we take is large enough to be representative of the population as a whole.

2 The sample should be taken at random. If we don't make sure that our sample is taken at random, our results may be biased. Suppose we are using quadrats to sample the vegetation in a field. We must not choose where to place the quadrats or we could end up with data which is unrepresentative.

You cannot ensure that a quadrat has been placed at random by throwing it over your shoulder, even if you have closed your eyes first. Always use the method described in the box.

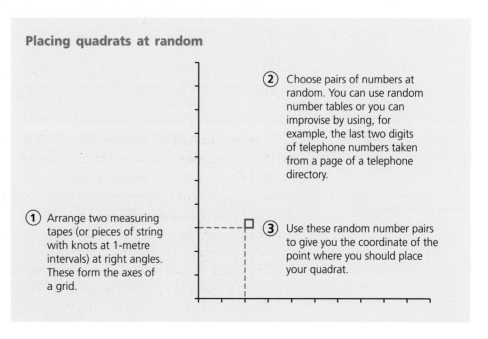

Placing quadrats at random

① Arrange two measuring tapes (or pieces of string with knots at 1-metre intervals) at right angles. These form the axes of a grid.

② Choose pairs of numbers at random. You can use random number tables or you can improvise by using, for example, the last two digits of telephone numbers taken from a page of a telephone directory.

③ Use these random number pairs to give you the coordinate of the point where you should place your quadrat.

Table 8.1 summarises how quadrats, point quadrats and transects may be used in sampling plants; although they can, of course, be used in investigating the distribution of other organisms.

Technique	What is it?	What is it used for?	How is it used?
Quadrat	A sample area marked out so that the organisms it contains can be studied.	Usually used to study the distribution of plants in a fairly uniform area.	Quadrats are placed at random and used to find population density, frequency or percentage cover (see below).
Point quadrat	This is a quadrat which is reduced in size to a single point.	Used to study the distribution of plants. Point quadrats are often used with transects.	A point frame is normally used. This consists of a group of 10 pins about 5 cm apart. By lowering the pins in turn and recording the number of hits on a particular species of plant, the frequency or percentage cover may be estimated.
Transect	A line along which samples are taken.	Studying areas where there is an environmental gradient and species vary.	A transect is positioned so that it follows the environmental gradient. A quadrat or point quadrat is often used at intervals along the transect to sample the area in more detail.

Table 8.1

There are three measures that are commonly used to describe the distribution of plants. These are:

1 **Population density.** This is the number of individual plants of a particular species in a given area. If you wanted to know the population density of dandelions in a field, you would place a number of quadrats at random and count the number of dandelion plants in each. From the information you collected, you could calculate the population density of dandelions in plants per square metre.

2 **Frequency.** Frequency is based on the number of quadrats, either ordinary quadrats or point quadrats, in which a species occurs. Suppose we use point quadrats to measure the frequency of dandelions. The pins on the point frame are lowered, and the number of hits is recorded. If 3 out of 10 pins hit dandelion plants, then the frequency is 30%.

3 **Percentage cover.** Percentage cover measures the proportion of ground in a quadrat occupied by a particular species.

4.2 Mark-release-recapture

This is a way of estimating the size of a population of animals that move around quite a lot. It relies on capturing a number of animals, marking them and releasing them. Some time later a second sample is captured and the numbers of marked and unmarked animals in this sample are recorded. From these data, the size of the population can be estimated.

If you have to calculate the size of a population using the mark-release-recapture method, don't try to learn a formula — it is far too easy to get it wrong. Just remember: **proportion of marked animals in sample = proportion of marked animals in population.**

Example

Number of mice captured, marked and released	40
Number of marked mice in second sample	15
Total number of mice in second sample	45

$$\textbf{proportion of marked mice in sample} = \textbf{proportion of marked mice in population}$$

$$\text{So,} \quad \frac{\textbf{number of marked mice in sample}}{\textbf{total number of mice in sample}} = \frac{\textbf{number of marked mice in population}}{\textbf{total number of mice in population}}$$

$$\frac{15}{45} = \frac{40}{\text{total population}}$$

$$\text{total population} = \frac{40 \times 45}{15}$$

$$= 120$$

The mark-release-recapture method only gives us an estimate of the number of organisms in a population. It involves a number of assumptions. These are:

- The number of animals in the population does not change between the time that the first sample is trapped and marked and the time that the second sample is trapped. This means that no new animals enter the population from births or immigration and no existing animals leave it as a result of death or emigration.
- When the marked animals are released, they must mix randomly in the population.
- Marking does not affect the animal in any way. It does not, for example, cause it distress or make it more conspicuous so that it is more likely to be captured by a predator.

5 *The diversity of living organisms*

When you have finished revising this topic, you should:

- be able to explain the meaning of the term diversity
- be able to explain what is meant by the terms succession and zonation
- be able to describe one example of succession
- know the general features which are associated with successional change
- be able to describe one example of zonation
- know the general features which are associated with zonation

5.1 Diversity

Diversity is a way of describing the relationship between the number of individual organisms and the number of species in a community. Sand dunes often have large numbers of marram grass plants growing on them and very

little else. They have a low diversity. On the other hand, tropical rain forests have enormous numbers of different species. They have a very high diversity. It is important to appreciate that diversity is more than just the number of species present. This can be illustrated by the information in Table 8.2. It shows the birds visiting bird tables over a period of one hour in two different gardens.

Species	Garden A	Garden B
House sparrow	23	12
Starling	3	4
Blue tit	1	16
Great tit	1	6
Greenfinch	2	11
Chaffinch	1	3

Table 8.2

Both gardens have the same number of species, but in garden **A**, almost all the birds visiting the bird table are house sparrows. Common sense tells us that there is a much higher diversity of birds in garden **B**.

5.2 Succession

Over a period of time, the different species of organisms in a community are gradually replaced by others. This is succession. Think about the soil that accumulates at the base of a wall along the side of a concrete yard. It is colonised by small plants which are adapted to these conditions. These are known as pioneer species. If left undisturbed, many of these will die. Leaves and stems drop off and decompose, adding humus to the soil. Conditions will now be suitable for other species of plant which will gradually replace the original pioneers. If left undisturbed for long enough, a climax community is eventually established. In Britain, this is generally woodland of some sort. A climax community is the final stage in an ecological succession.

Do not confuse ecological succession with evolutionary selection. In succession, one species is replaced by another; in selection, one form changes into another. Succession is a relatively short-term ecological process. Selection may take many millions of years and involves evolutionary change.

Some terms used in describing succession

Pioneer community
The organisms that first colonise an area.

Climax community
The final stage in an ecological succession. The type of community formed depends to a large extent on the climate. In much of the south of Britain it is oak woodland; further north it is likely to be pine woodland. Climax communities are stable and change very little.

Sere
A sere is a complete succession from pioneer community to climax community, and a seral stage is a particular stage in this succession with its own distinctive community of organisms.

Primary succession

This is a succession in which the pioneer organisms colonise a surface such as bare rock or sand dunes on which nothing had been growing before.

Secondary succession

In contrast to primary succession, secondary succession is where pioneer species colonise an area where there has already been a community of living organisms but these have now been removed. Examples are the bare ground resulting from clearing forest or ploughing farmland.

5.3 Sand dune succession

In almost any area where you might have undertaken fieldwork, it is possible to find an example of succession. The sand dunes found in many areas round our coasts provide a particularly good example. Figure 8.10 summarises the process of succession in sand dunes from the initial colonisation of bare sand to the establishment of a climax community.

Getting a good mark for a question about succession means bringing in A-level information. Don't just write about small plants being replaced by bushes and then by trees. Use the information in this sheet and include some real A-level biology.

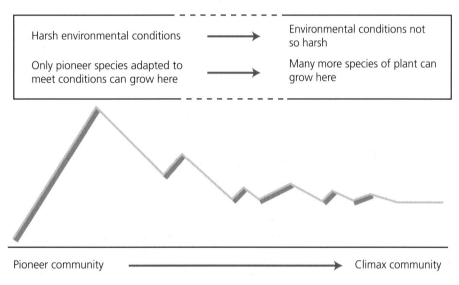

| Harsh environmental conditions | → | Environmental conditions not so harsh |
| Only pioneer species adapted to meet conditions can grow here | → | Many more species of plant can grow here |

Pioneer community ⟶ Climax community

• **Abiotic conditions**

Sand is easily blown by the wind. It is low in humus and holds little water. Levels of nitrates and other soil nutrients are low.

Sand is stabilised and does not blow about. Humus in soil holds much more moisture. Higher concentration of nitrates and other soil nutrients.

• **Plants**

The pioneer plants must be adapted to the harsh environmental conditions. Few species found because few species are adapted to these conditions. Many pioneer species produce large numbers of wind-dispersed spores or seeds.

Environmental conditions are far less harsh and more species can grow here. Larger, woody species become established. Many have large, animal-dispersed fruits.

• **Animals**

A few species of small plants means little variety of food or ecological niches for animals.

Many species of plants of differing size means greater variety of food and more ecological niches for animals.

Figure 8.10 Succession in sand dunes

5.4 Zonation

The abiotic conditions in a particular area are not constant. They vary, and this variation affects the distribution of living organisms. Zonation is the variation in the distribution of living organisms brought about by differences in environmental conditions. Good examples of zonation are provided by the distribution of seaweeds on a rocky shore and the effect of altitude on the distribution of mountain vegetation. Figure 8.11 summarises the process of zonation on a rocky seashore.

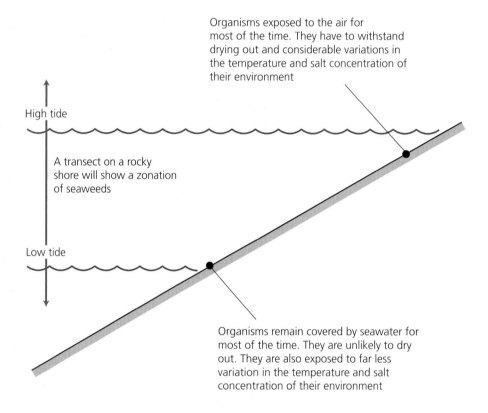

Figure 8.11 Zonation on a rocky seashore

6 | *Interactions between organisms*

When you have finished revising this topic, you should be able to:

- describe the following interactions:
  - competition
  - parasitism
  - mutualism
- explain the difference between intraspecific competition and interspecific competition
- interpret data relating to competition
- explain how parasites are adapted to their way of life
- explain how mutualism benefits the organisms concerned

6.1 Introduction

Living organisms can interact with one another in different ways. These interactions may involve benefits to both of the organisms concerned or there may be an advantage to just one of them. Figure 8.12 summarises some of these relationships.

Competition

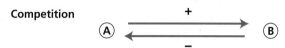

Organisms may compete for the same resource. In particular circumstances, one species will usually be better adapted than the other. This species will survive while the other will decrease in number.

Parasitism

In a parasitic relationship, the parasite gains a nutritional advantage; the host is at a disadvantage. In parasites that are well adapted to their way of life, the disadvantage to the host is very small.

Mutualism

Both organisms in a mutualistic relationship gain a nutritional advantage.

Figure 8.12 Some of the relationships which exist between organisms of different species

6.2 Competition

Competition is the term given to a relationship between organisms that require the same resource.

Intraspecific competition is competition between organisms belonging to the same species. Nettles growing in a clump will compete with each other for water and mineral ions such as nitrates and phosphates; the aphids feeding on them will compete with each other for the food that the nettles supply.

Interspecific competition is competition between organisms of different species; two-spot lady birds and seven-spot ladybirds both feed on the aphids found on nettle plants. They will compete with each other.

> **Interpreting data about competition**
>
> Questions about competition often require you to interpret data. There are two questions which you need to ask yourself before you answer.

When you write about competition, use the right language. Organisms 'compete' for resources. They do not 'fight' for them.

Don't confuse interspecific and intraspecific. Intercity trains go between different cities; interspecific competition is competition between different species.

1   Is this question about intraspecific competition or interspecific competition?

2   What is the resource for which the organisms concerned are competing? Let us look at a simple example of elephants competing with each other during the dry season. It is unlikely that they will be competing for space — there will surely be enough room for another elephant. It will not be oxygen. They all need oxygen for respiration but, again, it will not be in short supply. In this case the resource for which they are competing is more likely to be food or water. Table 8.3 lists some of the resources for which animals and plants frequently compete.

Animals	
Food	This is the resource for which animals most frequently compete.
Space	Not really important except in sessile animals such as barnacles which attach themselves to rocks and remain there permanently.
Breeding	There may be competition between animals of the same species for mates and breeding sites.
Plants	
Mineral ions	The mineral ions in the soil which are essential for plant growth are often in limited supply. Competition for these is often a factor which limits plant growth.
Water	Plants take up large amounts of water from the soil in the process of transpiration. In many ecosystems, water is often in short supply.
Light	Light is essential for photosynthesis. The shade cast by some plants often limits others.
Space	Competition for space is more important for plants, as they are unable to move to other areas.

Table 8.3

6.3 Parasitism

A parasite may be defined as an organism which lives in or on a host organism. The parasite gains by getting its food from the host. The host is clearly at a disadvantage.

Many different species live as parasites but they have similar adaptations to their way of life. These include:

- **Ways of attachment to their hosts.** A tapeworm has a ring of hooks and a number of suckers with which it can gain a grip on the intestine wall; head lice have claws with which they grasp the hair of their human hosts.
- **Various ways of avoiding or overcoming the host defence mechanisms.** Good examples of this are seen in parasites that live in the blood system or the gut of a mammal. Trypanosomes coat themselves in protein from their hosts so are not attacked by the immune system; the cuticle covering the body of a tapeworm is resistant to digestive enzymes.

If your specification requires a knowledge of parasitism, learn these five adaptations and make sure you know how they apply to the organisms listed in the specification.

- **Development of reproductive organs.** The chances of a parasite successfully completing its life cycle and infecting a new host are very remote. By producing large numbers of offspring, these chances are improved.
- **A complex life cycle.** This again improves the chances of a parasite infecting a new host.
- **The reduction of body systems other than those involved with reproduction.** A tapeworm, for example, living in the gut of its host has no need for a digestive system. It only needs to absorb the products of its host's digestion.

6.4 Mutualism

This is a relationship between two species of organism in which both gain a nutritional advantage. Table 8.4 shows two examples of mutualism which are found in A-level specifications.

Example	Advantages to organisms concerned
Cellulose digesting bacteria in rumen of a cow	The cow provides microorganisms with a constant food supply rich in cellulose. Microorganisms provide the cow with fatty acids from digestion of cellulose.
Nitrogen-fixing bacteria in root nodule of a leguminous plant	The leguminous plant provides bacteria with carbohydrate. The bacteria pass ammonium ions direct to the plant.

Table 8.4

> When you write about mutualism, try to identify the specific nutritional advantage gained by both organisms. Do not give general points like 'bacteria having a safe place to live' or 'somewhere safe from predators'.

7 *Populations and population biology*

When you have finished revising this topic, you should:

- be able to describe a population growth curve and explain the different stages in the growth of a population
- be able to calculate the rate of growth of a population from appropriate data
- be able to describe and explain the shape of curves showing predator and prey populations
- understand what is meant by biological control
- be able to explain the advantages of biological control over other methods of pest control and describe precautions which should be taken before releasing a control organism

7.1 Population growth curves

When an organism is introduced and successfully establishes itself in a new area, its population tends to grow slowly at first. It then undergoes a period of increasingly rapid population growth before reaching an equilibrium in which its numbers remain more or less constant. Figure 8.13 is a graph showing the changes which occur in a population of yeast cells which are growing in a sucrose solution.

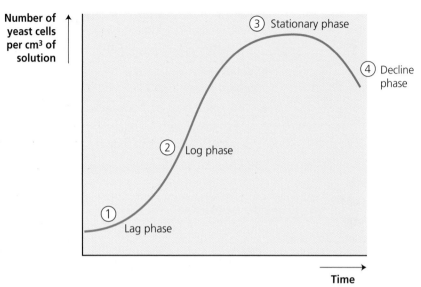

Figure 8.13 *A population growth curve for yeast cells growing in a sucrose solution*

The curve plotted on this graph shows four distinct stages or phases. These are:

1 **Lag phase.** At first, the increase in population of the yeast cells is very slow. They increase in size but not in number. During this stage, genes are switched on and enzymes are produced.

2 **Log phase.** This is a period of rapid population growth. Nutrients are plentiful and waste products have not built up to harmful levels.

3 **Stationary phase.** The population has reached equilibrium. The number of new cells added to the population is more or less balanced by the number of cells which are dying. A shortage of food and a build up of toxic waste products may limit the population.

4 **Decline phase.** In the final stage, the population of living cells starts to decrease. This may be due to a shortage of food or of a particular nutrient, or it may be due to the build up of toxic waste products such as ethanol.

A similar curve could be drawn for other organisms such as the rabbit population of Australia. In this case, explanations of the different phases may be a little different from those above. The lag phase, for example, will not involve the activation of genes and the production of different enzymes. It is more likely to be concerned with changes in behaviour as the rabbits adapt to their new environment. Similarly, the factors affecting the stationary phase may differ. These limiting factors will probably involve a shortage of food, but they may also include increased likelihood of disease.

Calculating rate

Many students find difficulties calculating rates. Here is a worked example concerned with rate of population growth.

Table 8.5 shows the number of yeast cells in 1 mm^3 of sucrose solution over a 5-hour period.

Calculate the mean rate of increase of the yeast population between 0 and 3 hours.

Time (hours)	Number of yeast cells in 1 mm³ of sucrose solution
0	1.8×10^3
1	2.1×10^3
2	2.4×10^3
3	2.7×10^3
4	2.9×10^3
5	3.1×10^3

Figure 8.5
A population growth

All you have to do in calculating rate in an example like this is to divide the increase in population by the time over which it occurs.

$$\text{rate of population growth} = \frac{\text{increase in the number of yeast cells}}{\text{time}}$$

$$= \frac{(2.7 \times 10^3) - (1.8 \times 10^3)}{3 - 0}$$

$$= \frac{0.9 \times 10^3}{3} = 0.3 \times 10^3 \text{ cells per hour}$$

Don't forget to give the units in rate calculations. You will need to give these if you want full marks.

7.2 Predator and prey

The population of a predator will be affected by the population of the prey on which it feeds. It has been suggested that this gives rise to regular cycles in prey and predator populations. A low prey population means that there will be little food for the predators. The prey population will therefore increase, providing more food for the predators. The predator population increases in turn and reduces the population of its prey. The result of this is that both predator and prey numbers oscillate as shown in Figure 8.14. There is some debate over whether this is really what happens, because prey populations show regular fluctuations even in the absence of predators.

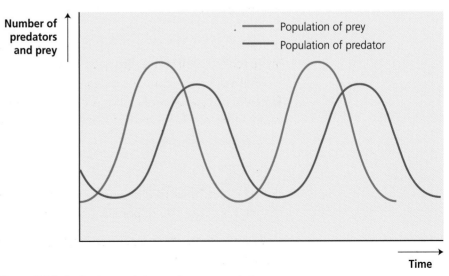

Figure 8.14 Cycles in predator and prey populations

Illustrate answers about biological control with good examples. Do not write about gardeners releasing ladybirds on rose bushes to control aphids or farmers keeping cats to kill mice.

Organisms used in the biological control of insects which affect plants will not eat the crop if numbers of pests decrease. They will probably die because they have run out of food.

7.3 Biological control

Species often become pests when they are introduced into a new environment. In the absence of their natural predators and parasites, their numbers increase rapidly. Biological control works on the principle that introducing a predator or parasite of the pest will reduce its population to a level where it no longer causes enough damage to be economically important. Examples of biological control include the use of the virus which causes myxomatosis to control rabbits, and the parasitic wasp, *Encarsia*, which has been used successfully to control populations of whitefly in large glasshouses.

Before an organism is used for biological control, care has to be taken that:
- It will be able to survive and establish itself in the new conditions.
- It must eat or parasitise only the pest. It must not affect other organisms.

7.4 The advantages of biological control

Biological control has a number of advantages over the use of chemical pesticides. These include the following:
- Once established, the parasite or predator used for biological control does not have to be reintroduced. Chemical pesticides have to be applied frequently.
- Pests do not usually evolve resistance to parasites and predators. Many chemical pesticides can only be used for a relatively short time, because the pest rapidly evolves resistance to them.
- Organisms used for programmes of biological control are very specific and only affect the pest. Pesticides frequently affect other harmless or even beneficial organisms.

8 Human influences on the environment

When you have finished revising this topic, you should:
- understand how energy resources can be managed in a sustainable manner
- be able to describe the processes of deforestation and desertification and the consequences of these processes
- be able to define the term pollution and describe different forms of pollution
- understand the importance of conservation

8.1 Energy sources

It is important for biologists to understand how energy resources can be managed in a sustainable manner. This involves conserving energy resources, using the resources more efficiently and seeking alternative (renewable) resources.

Non-renewable energy resources are fossil fuels, such as coal and oil. They are relatively cheap and readily available, but supplies are limited and the burning of these fuels causes atmospheric pollution.

Renewable energy resources are generally more complex to produce in a useable form and so more expensive than fossil fuels. However, supplies of some of these resources are, theoretically, unlimited. Examples include: fast-growing biomass (e.g. willow); gasohol from sugar; and biogas from domestic and agricultural wastes.

8.2 Deforestation and desertification

Deforestation is the process whereby trees are removed to provide timber and fuel, and to supply agricultural land for crops and animals. Excessive deforestation results in soil erosion, loss of biodiversity and global warming (via the greenhouse effect).

Desertification is usually caused by grazing animals and fuel wood gathering, reducing the levels of vegetation in dry areas. This leads to soil erosion and salinisation (an increase in the salt concentration in the soil). This results in the gradual spread of desert and a reduction in the availability of fertile land.

8.3 Pollution

Pollution may be defined as a change in the abiotic or biotic characteristics of the environment as a result of human activities, such as pollution of the atmosphere and water supplies.

Atmospheric pollution may be due to the increased production of sulphur dioxide and oxides of nitrogen during the burning of fossil fuels. This leads to acid rain, resulting in damage to aquatic and terrestrial ecosystems, and the erosion of buildings. Carbon dioxide, nitrous oxide, methane, ozone and chlorofluorocarbons (CFCs) in the atmosphere contribute to the greenhouse effect and global warming; resulting in rising sea levels and a disruption of ecosystems.

Water pollution may be caused by a number of factors, such as the leaching of fertilisers into rivers and lakes, leading to **eutrophication**. Eutrophication is the decrease in biodiversity resulting from the pollution of a river or lake.

8.4 Conservation

Conservation is the process of maintaining an ecosystem in order to retain maximum species diversity. This involves careful use of the Earth's natural resources in order to avoid the gradual destruction of the environment by human activities. It also involves preservation of natural habitats and an awareness of the dangers of pollution.

Conservation of resources is essential for economic and survival reasons, also for ethical reasons — we have a duty to protect our planet and to pass on sufficient resources to future generations. The key resources required for human life are:
- food and water
- land
- energy
- minerals and other raw materials
- other species

Water treatment processes are used to recycle water. Sewage treatment results in relatively clean water than can be released into rivers and lakes. Further water treatment converts this water into drinking water. Active conservation can reclaim land; for example, spoil heaps produced by mining activities can be landscaped and covered with topsoil. As long as they are fee from toxins, these areas can then be used as farmland or amenity areas.

Some resources, such as fossil fuels, cannot be recycled. Others, such as paper, glass and plastic, can be recycled relatively easily. The recycling of glass and plastic actually uses less energy than making these resources from scratch.

Biological conservation maintains biodiversity. Living resources are used by humans as a source of food, in medicines and in biotechnology. Conserving species means conserving habitats; many species are under threat because their natural habitats are shrinking due to:

- expansion of human populations
- over-exploitation of resources, e.g. deforestation.

Several measures are in place to promote habitat conservation in the UK.

- **National parks** set up to conserve the natural beauty of an area and protect the organisms that live there.
- **Sites of Special Scientific Interest** (SSSIs) smaller areas that are particularly important because of their wildlife.
- **Environmentally Sensitive Areas** (ESAs) areas of significant ecological interest, where farmers are encouraged to adopt environmentally friendly land management practices.
- **Wildlife reserves** to maintain and protect areas for rare species to live and breed safely.

Species can also be conserved outside their natural habitats, such as in zoos and botanical gardens.